ENGAA Past Paper Worked Solutions

UniAdmissions

ENGAA Past Paper

Worked Solutions

Peter Stephenson

Dr. Rohan Agarwal

UniAdmissions

About the Authors

Peter is an Oxbridge Admissions tutor at *UniAdmissions*. He is currently in his 4th year of studying Physics at St. Catherine's College, Oxford. Peter achieved a distinction in each of his first three years and has been awarded an ATV Scholarship. Next year he is starting a PhD at Imperial College, focussing on plasma atmospheres around comets.

Outside of academia, Peter enjoys playing rugby and escaping into the countryside to climb some very tall rocks.

Rohan is the **Director of Operations** at *UniAdmissions* and is responsible for its technical and commercial arms. He graduated from Gonville and Caius College, Cambridge and is a fully qualified doctor. Over the last five years, he has tutored hundreds of successful Oxbridge and Medical applicants. He has also authored ten books on admissions tests and interviews.

Rohan has taught physiology to undergraduates and interviewed medical school applicants for Cambridge. He has published research on bone physiology and writes education articles for the Independent and Huffington Post. In his spare time, Rohan enjoys playing the piano and table tennis.

THE BASICS

Hundreds of students take the ENGAA exam each year. These exam papers are then released online to help future students prepare for the exam. Since the ENGAA is such a new exam, past papers have become an invaluable resource in any student's preparation.

Where can I get ENGAA Past Papers?

This book does not include ENGAA past paper questions because it would be over 500 pages long if it did! However, all ENGAA past papers since 2016 (including the specimen paper) are available for free from the official ENGAA website. To save you the hassle of downloading lots of files, we've put them all into one easy-to-access folder for you at **www.uniadmissions.co.uk/ENGAA-past-papers**.

How should I use ENGAA Past Papers?

ENGAA Past papers are one the best ways to prepare for the ENGAA. Careful use of them can dramatically boost your scores in a short period of time. The way you use them will depend on your learning style and how much time you have until the exam date but here are some general pointers:

➢ Four to eight weeks of preparation is usually sufficient for most students.

➢ Make sure you are completely comfortable with the ENGAA syllabus before attempting past papers – they are a scare resource and you shouldn't 'waste them' if you're not fully prepared to take them.

➢ Its best to start working through practice questions before tackling full papers under time conditions.

➢ You can find two additional mock papers in the *ENGAA Practice Papers* Book (flick to the back to get a free copy).

How should I prepare for the ENGAA?

Although this is a cliché, the best way to prepare for the exam is to start early – ideally by September at the latest. If you're organised, you can follow the schema below:

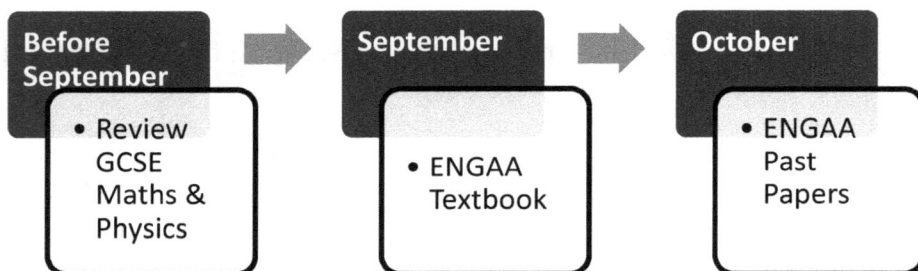

Before September		September		October
• Review GCSE Maths & Physics	⟹	• ENGAA Textbook	⟹	• ENGAA Past Papers

This paradigm allows you to minimise gaps in your knowledge before you start practicing with ENGAA style questions in a textbook. In general, aim to get a textbook that has lots of practice questions e.g. *The Ultimate ENGAA Guide* (**www.uniadmissions.co.uk/ENGAA-book**) – this allows you to rapidly identify any weaknesses that you might have e.g. Newtonian mechanics, simultaneous equations etc.

You are strongly advised to get a copy of *'The Ultimate ENGAA Guide'* which has 250 practice questions– you can get a free copy by following the instructions at the back of this book.

Finally, it's then time to move onto past papers; you should endeavour to do every past paper at least twice before your test date.

If you find that you've exhausted all past papers, there are an additional two mock papers available in the *ENGAA Practice Papers* Book (flick to the back to get a free copy).

.

How should I use this book?

This book is designed to accelerate your learning from ENGAA past papers. Avoid the urge to have this book open alongside a past paper you're seeing for the first time. The ENGAA is difficult because of the intense time pressure it puts you under – the best way of replicating this is by doing past papers under strict exam conditions (no half measures!). Don't start out by doing past papers (see previous page) as this 'wastes' papers.

Once you've finished, take a break and then mark your answers. Then, review the questions that you got wrong followed by ones which you found tough/spent too much time on. This is the best way to learn and with practice, you should find yourself steadily improving. You should keep a track of your scores on the next page so you can track your progress.

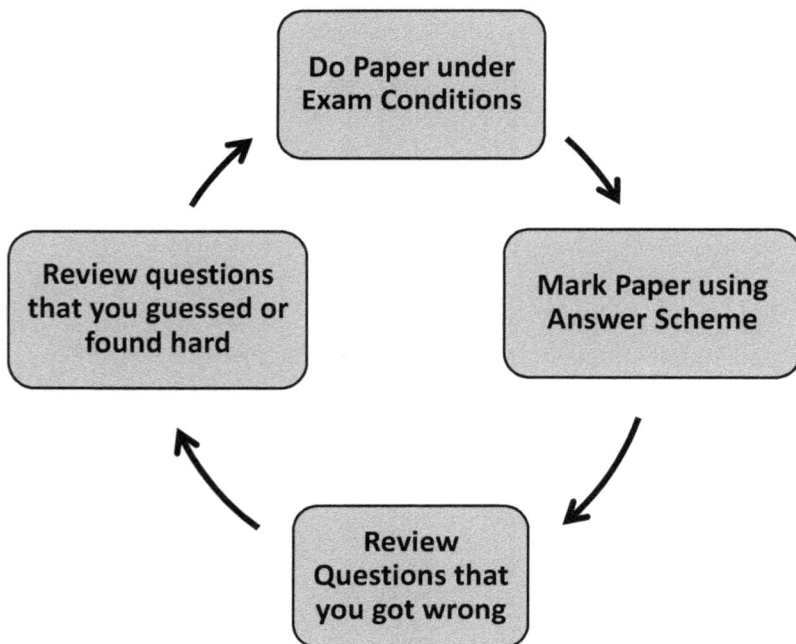

```
        ┌──────────────────────┐
        │   Do Paper under     │
        │  Exam Conditions     │
        └──────────────────────┘
   ↗                              ↘
┌──────────────────┐      ┌──────────────────┐
│ Review questions │      │ Mark Paper using │
│ that you guessed │      │  Answer Scheme   │
│   or found hard  │      └──────────────────┘
└──────────────────┘              ↙
        ↖      ┌──────────────────┐
               │     Review       │
               │ Questions that   │
               │  you got wrong   │
               └──────────────────┘
```

Scoring Tables

Use these to keep a record of your scores – you can then easily see which paper you should attempt next (always the one with the lowest score).

SECTION 1	1st Attempt	2nd Attempt	3rd Attempt
Specimen			
2016			
2017			
Mock Paper A			
Mock Paper B			

SECTION 2	1st Attempt	2nd Attempt	3rd Attempt
Specimen			
2016			
2017			
Mock Paper A			
Mock Paper B			

SPECIMEN

Section 1

Question 1: A

Take each side of the square as length a. The removed semicircle has radius a/2. The area of the square is a^2 and the area removed is $\pi a^2/8$. Therefore,

$$100 = a^2 \left(1 - \frac{\pi}{8}\right)$$

Rearranging this gives:

$$a^2 = 800/(8 - \pi)$$

Taking the square-root gives:

$$a = 20\sqrt{\frac{2}{8 - \pi}}$$

Question 2: C

The net force is 300N upwards. F = ma. The parachutist has mass 60kg.

Therefore,

$$a = \frac{F}{m} = 5\text{ms}^{-2}$$

In the upwards direction

Question 3: E

The ratio 1:2 indicates RQ is half the length of PQ, i.e. 10cm.

Call the point on the perpendicular and the hypotenuse S.

The length of PR is $10\sqrt{5}$cm using Pythagoras theorem. The triangle PQR and PSQ are similar. The ratio of RQ:PR is equivalent to that of SQ:PQ. Therefore SQ $= 20 \times 10/(10\sqrt{5})$. Simplifying this gives $4\sqrt{5}$.

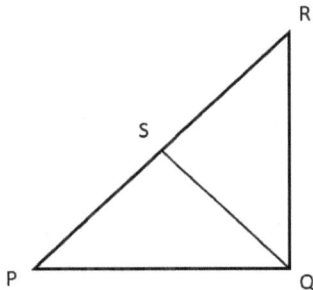

Question 4: C

Use the equation $c = f\lambda$. From the graph the wave does one whole oscillation in 2 seconds giving a frequency of $f = 0.5$Hz. The wavelength of the wave is 1.5cm. Multiplying the two gives $c = 0.75$cms^{-1}

Question 5: A

The length of the dotted line is x. First consider the distance from the centre of the base to one corner of the base. This is given by $\sqrt{(0.5)^2 + (0.5)^2}$. Now consider the triangle the has its hypotenuse along the dotted line. The other two sides are the vertical edge and the line from the centre of the base to the corner of the base. The vertical has length 1. Now use Pythagoras to find the length. $x^2 = 1^2 + 0.5^2 + 0.5^2 = 1.5$. Taking the square-root gives $\sqrt{3/2}$.

Question 6: B

The kinetic energy is $E = mv^2/2$. The momentum is given by $p = mv$. Divide the first equation by the second giving $5 = v/2$. Therefore $v = 10$ms^{-1}. Substituting this into the momentum equation gives 3kg

Question 7: D

The 12 hours span 360 degrees. Therefore, each hour is 30 degrees. At 9:45 the hour hand has moved ¾ of the way to 10. The answer is given by $30 \times 0.75 = 22.5$

Question 8: E

Consider the equation $P = IV$. Power has the unit Watt and Current has the unit of amps. Therefore voltage, or potential difference, is given by Watt per amps.

Question 9: B

The area of a triangle is given by $A = \frac{1}{2}bh = \frac{1}{2}(4 + \sqrt{2})(2 - \sqrt{2})$

Expanding the brackets gives $\frac{1}{2}(8 - 2\sqrt{2} - 2) = 3 - \sqrt{2}$

Question 10: D

In 24 hours source X has undergone 24/4.8=5 half-lives and source Y has undergone 24/8=3 half-lives. Therefore, the activity of X has decreased by a factor of $\left(\frac{1}{2}\right)^5$. This gives $320 \times \frac{1}{32} = 10$.

The activity of Y is $1/8^{\text{th}}$ of its initial value. i.e. 60Bq

Adding these two contributions gives 70Bq.

Question 11: D

The radius of the cylinder is r and the height of the cylinder is 2r. It has a volume of $\pi r^2 h = 2\pi r^3$. Dividing the area of the sphere by this gives 2/3.

Question 12: E

The potential energy lost is given by $E = mgh = 100 \times 10 \times 100 = 100\,000J$.

The distance travelled down the slope is 10 times the vertical distance dropped i.e. 1000m. For constant speed all the energy must be dissipated by resistive forces. The work done by the resistive force is $W = Fd$.

Therefore, $F = W/d = 100000/1000 = 100N$

Question 13: A

For $0 < x < 1$: $x^2, \sqrt{x}$ are both less than one. For the other three, want the denominator to be as small as possible to give the largest number. $1 + x < x$ rules out option C. In this range of x, $x < \sqrt{x}$. The largest option is $1/x$

Question 14: A

Assuming all the initial kinetic energy is converted into gpe,

$$KE = \frac{1}{2}mv^2 = mgh = GPE$$

Rearranging gives $h = v^2/(2g) = 144/(20) = 7.2m$

Question 15: C

The ratio AB:BC is equal to the ratio AD:DE. AD has length x cm.

Therefore, $\frac{4}{x} = \frac{x}{x+3}$. Rearranging this gives $4(x + 3) = x^2$

Thus, $x^2 - 4x - 12 = 0$

Factorising this gives $(x - 6)(x + 2) = 0$

The two possible solutions are $x = 6$ cm and $x = -2$cm. A negative length is un-physical, and this solution can be discarded. The length of DE is $x + 3 = 9$cm

Question 16: C

To bring the lorry to rest all the kinetic energy must be dissipated by the resistive forces. The initial energy is given by $\frac{1}{2}mv^2$. The resistive force must do work given by $W = Fd$ which is equal to the initial energy. Thus, $Fd = \frac{1}{2}mv^2$.

Rearrangement gives $d = mv^2/(2F)$

Question 17: C

As Q is 1.4 times its original value Q^2 is (1.4x1.4)=1.96 times its original value. Therefore, P is $1/1.96$ its original value. This is close to $1/2$ i.e. a 50% decrease. However, $1/1.96$ is larger than $1/2$. Therefore, it has decreased by a factor of close to but less than 50%. The answer is 49%.

Question 18: B

An alpha particle consists of 2 neutrons and two protons so alpha decay would decrease the value of the mass number by 4 and the proton number by 2. A beta decay ejects an electron from the nucleus by converting a neutron to a proton. This doesn't affect the mass number but increases the proton number by one.

From X to Y an alpha decay must occur, so P is equal to N-4. From Y to Z the mass number is unchanged, so it must be a beta decay. Thus, Q equals R-1.

Question 19: D

$x \propto z^2$ and $y \propto 1/z^3$. This means $x^3 \propto z^6$ and $y^2 \propto 1/z^6$

Inverting the second equation yields $z^6 \propto 1/y^2$. Combining this with the first equation gives $x^3 \propto 1/y^2$

Question 20: B

For the pulse to travel to the foetus and back it must travel 20cm. It travels at a speed of $500ms^{-1}$. The time taken is given by $t = 0.2/500 = 0.4ms$.

Question 21: C

Call the distance QX x. Therefore, the distance PX is 6x and the distance XR is $\frac{3}{2}x$.

The distance PR is $\frac{15}{2}x$. As M is the midpoint of PR the distance PM is $\frac{15}{4}x$. The distance MX is $PX - MX = 6x - \frac{15}{4}x = \frac{9}{4}x$

Therefore, $\frac{QX}{MX} = \frac{x}{(9x)/4} = \frac{4}{9}$.

Question 22: B

Constant acceleration gives a straight line on a velocity-time graph, so we can discard **R** and **S**. Graph **P** has an acceleration of $10/24ms^{-2} \neq 2.4ms^{-2}$. The final graph **Q** reaches a velocity of $58ms^{-1}$. The change in velocity is therefore $48ms^{-1}$ over 20s i.e. an acceleration of $2.4ms^{-2}$.

Question 23: D

Move all the terms to the left-hand side: $x^2 + 2x - 8 \geq 0$

Factorising this gives: $(x + 4)(x - 2) \geq 0$

Therefore $x^2 + 2x - 8$ changes sign at $x = -4$ and $x = 2$. Note that the x^2 term is positive, so the graph is U-shaped. It is therefore greater than or equal to 0 when $x \geq 2$ or when $x \leq -4$.

Question 24: D

Fission is when a nucleus splits into two parts- not gamma decay. A half-life is the time taken for half of the substance to decay but each nuclei decays instantaneously at a random time- B is false. The number of neutrons is the mass number – the atomic number- C is False. Under beta decay a neutron in the nucleus becomes a proton and an electron. The electron is ejected so the number of particles in the nucleus is conserved. An alpha particle is made up of two protons and two neutrons which are ejected from the nucleus. No neutrons are converted to protons. D is the true statement.

Question 25: A

The area of a cylinder is given by $\pi r^2 h$. The surface area is $2\pi r^2 + 2\pi rh$

$$\pi r^2 h = 2\pi r^2 + 2\pi rh$$

$$rh = 2r + 2h$$

$$h(r - 2) = 2r$$

$$h = \frac{2r}{r - 2}$$

Question 26: D

The two resistances in series add. The total resistance of the circuit is $R_1 + R_2$

Using Ohm's Law gives the current through the circuit as $I = \frac{V}{R} = \frac{V}{R_1 + R_2}$

This is the current that flows through each resistor and the power dissipated by the first resistor is given by $P = IV = I^2 R = \frac{V^2 R_1}{(R_1 + R_2)^2}$.

Question 27: B

From **1** to **2** the square has been rotated by 90 degrees clockwise. It has then been reflected in $y = x$. To transform **3** to **1** the square must be reflected in the y axis.

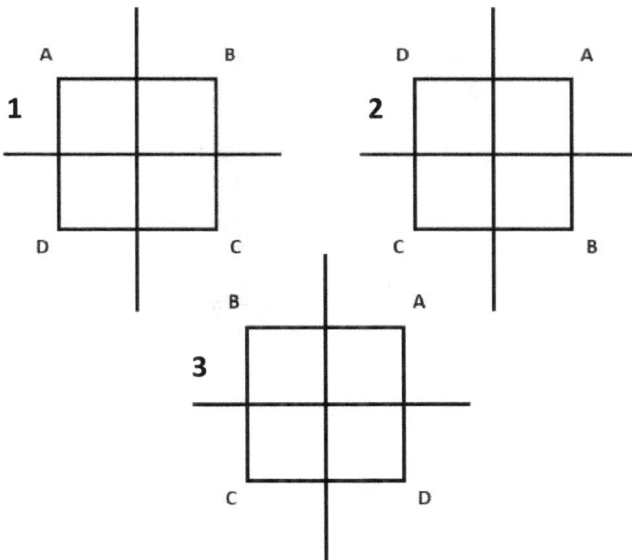

Question 28: B

The speed of sound is around $340ms^{-1}$ so P is not true. The distance between the point X and Y corresponds to twice the amplitude i.e. the peak to peak distance. Therefore, the amplitude is 2.5mm. The wavelength cannot be deduced from the information given as the speed is not given. The frequency is $f = \frac{1}{T} = 5Hz$. The only statement that can be deduced is S.

Question 29: D

$$\frac{x^2 - 4}{x^2 - 2x} = \frac{(x + 2)(x - 2)}{x(x - 2)} = \frac{x + 2}{x}$$

Question 30: A

The acceleration of the car is uniform, so the force is constant. Use $F = ma$ with an acceleration of $a = \frac{10}{5} = 2ms^{-2}$. This gives a net force of 2kN. The resistive force must therefore be $f = 3 - 2 = 1kN$

Question 31: F

$$a^x b^{2x} c^{3x} = (ab^2 c^3)^x = 2$$

$$\log 2 = x \log(ab^2 c^3)$$

$$x = \frac{\log(2)}{\log(ab^2 c^3)}$$

Question 32: C

Consider P initially moving from left to right and Q from right to left. In the collision momentum is conserved. The total initial momentum is given by

$p = 2 \times 3 - 5r = 6 - 5r$. The final momentum is given by $\frac{5r}{2} - 2$.

Equating these gives $6 - 5r = \frac{5r}{2} - 2$. Rearrangement gives $r = \frac{16}{15}$.

Question 33: D

A gives $\tan\left(\frac{3\pi}{4}\right) = -1$. B is $\log_{10} 100 = 2 \log_{10} 10 = 2$. C is $\sin(\pi/2) = 1$ and $1^{10} = 1$. D gives $\log_2 10 > \log_2 8 = 3$ Therefore, D is larger than 3. $\sqrt{2} - 1 < 1$, so, the tenth power of this is very small and certainly less than 1. Thus, D is the largest.

Question 34: A

Initially, the parachutist should experience an increasing air resistance until they reach terminal velocity when it becomes constant. When the parachute is opened the air, resistance will increase substantially. Then as the parachutist decelerates the air resistance will then decrease until the new terminal velocity is reached. As the parachutist experiences no net force at terminal velocity the air resistance must be the same as in the previous period of terminal velocity. The only graph that satisfies this is A.

Question 35: E

Call $y = 2^x$. Then substitute this into the original equation, giving $y^2 - 8y + 15 = 0$. Factorising this gives $(y - 5)(y - 3) = 0$. This yields two solutions: $2^x = 5$ and $2^x = 3$. Taking logs of these gives solutions

$$x_1 = \frac{log_{10} 5}{log_{10} 2} \text{ and } x_2 = \frac{log_{10} 3}{log_{10} 2}$$

Summing these and factoring out the denominator gives:

$$x_1 + x_2 = \frac{1}{log_{10} 2} (log_{10} 5 + log_{10} 3) = \frac{log_{10} 15}{log_{10} 2}$$

Question 36: F

The first graph decreases at a decreasing rate. This could be either the acceleration as it reaches terminal velocity or the resultant force on the car. The gravitational energy does not change as the car travels horizontally. The second graph increase to a steady state. This could by air resistance or velocity which are both constant at terminal velocity. Finally, Z can only be the weight of the car which remains constant throughout. The kinetic energy increases as the car accelerates. The only option which satisfies all three is option F.

Question 37: E

The first inequality can be shown to be true by subtracting both a and b from both sides. For the second inequality move the 2ab to the left-hand side $a^2 + b^2 - 2ab = (a - b)^2 \geq 0$. This is true assuming both a and b and real numbers. The third inequality can be broken for a negative value of c. Therefore, the answer is E.

Question 38: C

Initially the frictional force balances the horizontal force applied. However, the frictional force cannot exceed μR so after a time the block will accelerate. As the horizontal force continues to increase and the frictional force remains constant the block will increasingly accelerate.

Question 39: A

The term $(-1)^n$ is equal to -1 when n is odd and +1 when n is even. This means the first six terms of the sequence are: $2, 1, 2, 1, 2, 1$. Therefore, when n is odd $a_n = 2$ and for n even $a_n = 1$. The summation to n=100 can be written as $2 \times 50 + 1 \times 50 = 150$.

Question 40: C

Require momentum to be conserved during the collision. The balls all have mass m. The initial momentum is 3m. Call the final velocity of the red ball v. Then the final momentum is $m + mv = 3m$. Divide by m and this yields $v = 2ms^{-1}$

Question 41: C

Roots are found where the graph $y = x^4 - 4x^3 + 4x^2 - 10 = 0$. To find how many points satisfy this consider the location of the stationary points. $\frac{dy}{dx} = 4x^3 - 12x^2 + 8x = 4x(x^2 - 3x + 2) = 4x(x - 1)(x - 2)$. Stationary points are found where this equals zero i.e. $x = 0, x = 1, x = 2$. Now we want to find whether these stationary points are located above or below the x axis. $y(0) = -10. y(1) = -9. y(2) = -10$. All 3 of these are located beneath the x axis so there are no roots between x=0 and x=2. However, when x is a large negative or positive value the x^4 term dominated so y is a large positive value. Therefore, it must intersect the x axis at 2 points, so it has two roots.

Question 42: B

All the kinetic energy of the rock is converted into gravitational potential energy, so we can write the equation $\frac{mv^2}{2} = mgh$. Cancelling the masses and inserting the velocity of 20ms^{-1} gives $gh = 200$. The only option which satisfies this equation is option B.

Question 43: F

Consider a point in the centre of the base PQRS. Find the distance of this point from point P using Pythagoras theorem: $x^2 = 5^2 + 5^2$. $x = 5\sqrt{2}$cm. Now form a right-angled triangle using the point at the centre of the base, P and T. The desired angle is the one at point P. We have the adjacent, with length $5\sqrt{2}$cm, and the hypotenuse, 12cm. Using $\cos\theta = \frac{A}{H} = \frac{5\sqrt{2}}{12}$ gives the answer as F.

Question 44: A

The ball initially only has gravitational potential energy which as it falls is all converted into kinetic. During the bounce half of the energy is dissipated. The remaining energy is then all converted into gravitational potential energy as the ball rises. As GPE = mgh a 50% decrease in the energy causes the height of the bounce to decrease by 50% as the gravitational field strength and mass are constant. After the first bounce the ball reaches 8m, after the second 4m, after the 3rd 2m and after the 4th 1m. Therefore, the answer is A.

Question 45: D

For variables x and y, a straight line has the form $y = mx + c$. For a graph of $\log y$ and $\log x$ the relation should look like $\log y = m \log x + c$ with m and c both constants. The first gives $\log y = \frac{x \log a}{6}$. B gives $\log y = x \log b + \log a$. C gives $2 \log y = \log(a + x^b)$. D gives $\log y = b \log x + \log a$. E gives $\log y = \frac{a}{x} \log b$

The only one of the correct form is D.

Question 46: E

If the elevator was moving at constant velocity the motion would have no effect on the scale reading. If the elevator was moving downward with a decreasing speed the scales would have to exert a net force upwards on the man. This would cause the scale reading to be greater than the weight of the man. If the elevator were moving upwards with increasing speed the scale would have to exert a force great enough to accelerate the man upwards, so the reading would be larger than the man's weight. If the elevator is moving upwards with decreasing speed, the resultant force is downwards so the normal force of the scales is smaller.

Question 47: D

For real distinct roots in a quadratic equation we require the discriminant $b^2 - 4ac > 0$. For the given equation this is $(a - 2)^2 + 8a > 0$. Expanding the brackets yields $a^2 + 4a + 4 = (a + 2)^2 > 0$. If a is a real number, this is satisfied when $a \neq -2$.

Question 48: B

Constant acceleration of $a = -1.5\text{ms}^{-1}$. Can use SUVAT equations. Known variables are a, u and v to find s. Choose equation $v^2 = u^2 + 2as$. Rearranging this gives $s = \frac{1}{2a}(v^2 - u^2)$. Insert $u = 12\text{ms}^{-1}$ and $v = 0\text{ms}^{-1}$. This gives $s = 48\text{m}$

Question 49: C

To be able to construct a triangle the longest side must be shorter than the sum of the other two, so the longest side must be less than 6cm long. Therefore, there are only 3 possible constructions with sides of integer length: 5cm, 4cm and 3cm; 5cm, 5cm and 2cm; and 4cm, 4cm and 4cm.

Question 50: C

The vertical component of the tension is the angled rope must balance the weight of the particle. Resolving vertically gives $T_1 \cos(60) = T_1/2 = 5N$. This gives $T_1 = 10N$. The forces must also balance horizontally. The horizontal component of the tension in the angled rope is $T_1 \sin(60) = 5\sqrt{3}N$. This is directly balance by the horizontal rope, so the answer is C.

Question 51: F

If the angle at R is 90 degrees, the hypotenuse must be the opposite side i.e. PQ. Therefore, the length QR is given by $QR^2 = PQ^2 - PR^2$. To minimise QR we must minimise PQ whilst maximising PR. The smallest possible value of PQ is 3.5cm and the maximum of PR is 2.5cm. Inserting these gives $QR^2 = 6$. Thus, the minimum value of QR is $\sqrt{6}$

Question 52: D

Power is given by energy per unit time. The change in gravitational energy per unit time is given by $mg\frac{\Delta h}{\Delta t}$. Take the gradient of the graph when h=10m. $\frac{\Delta h}{\Delta t} = 15/30 = 0.5$. Using this gives $20 \times 10 \times 0.5 = 100W$

Question 53: B

In the range $0 \le x \le \pi$, $-1 \le \tan x \le 1$ is satisfied when $x \le \pi/4$ or $x \ge \frac{3\pi}{4}$.

The inequality $\sin y \ge 0.5$ is satisfied for $\pi/6 \le y \le 5\pi/6$. Substituting $y = 2x$ gives $\frac{\pi}{12} \le x \le \frac{5\pi}{12}$

These are both satisfied in the range $\frac{\pi}{12} \le x \le \frac{\pi}{4}$. Therefore, the interval has length $\pi/6$.

Question 54: A

All the carriages must accelerate at the same rate. The total mass of the system is 30 000kg. $a = F/m = 0.5 \text{ms}^{-1}$. Consider now only the forces acting on the final carriage. There is only the tension T which causes an acceleration of 0.5ms^{-1}. This gives $T = ma = 5000 \times 0.5 = 2500 \text{N}$

Section 2

Question 1a: A

The block is moving down the slope, so the resistive force acts up the plane. The gravitational force will always act vertically downwards. The normal force from the plane on the block acts perpendicular to the plane. This gives the answer as A.

Question 1b: D

The block is accelerating down the plane, so the forces do not balance in this direction. The block is also accelerating in the vertical and horizontal directions. The only direction in which the block is not accelerating is perpendicular to the plane so the forces balance in this direction

Question 1c: D

The gravitational force must be resolved into components parallel and perpendicular to the plane. The component parallel to the plane is $mg \sin \alpha$ and the component perpendicular to the plane is $mg \cos \alpha$. The frictional force acting up the plane is $\mu N = \mu mg \cos \alpha$. This gives the acceleration down the plane as $a = \frac{F}{m} = g(\sin \alpha - \mu \cos \alpha)$. As it starts at rest and the acceleration is constant the velocity is $v = g(\sin \alpha - \mu \cos \alpha)t$.

Question 1d: D

As $t \to \infty$ the velocity of the block will increase until the drag force balances the force down the plane from the previous part. This means $kv = mg(\sin \alpha - \mu \cos \alpha)$ at large times. The terminal velocity is therefore $v = mg(\sin \alpha - \mu \cos \alpha)/k$

Question 2a: D

The resistance is given by $R = \frac{\rho L}{A}$ where ρ is the resistivity, L is the length of the wire and A is the cross-sectional area. The area is given by $A = \pi r^2 = \pi \times (1.5 \times 10^{-2})^2 = 2.25\pi \times 10^{-4} m^2$. Therefore, $\frac{R}{L} = \rho/A = (2.6/2.25\pi) \times 10^{-8} \times 10^4 = 0.37 \times 10^{-4} \Omega m^{-1} = 37 m\Omega km^{-1}$

Question 2b: E

The circuit is made up of the power station, two cables and an unknown part which is the city. The power outputted by the powers station generates a current $P = IV$. This gives $I = \frac{1 \times 10^9}{4 \times 10^5} = 2.5 \times 10^3$A. The resistance of each cable is 3.7Ω. Use $P = I^2R = 6.25 \times 10^6 \times 3.7 = 23 \times 10^6$W to find the power dissipated in each cable. As there are two cables the total dissipated power is 46MW.

Question 2c: B

$E = P\Delta t$. Watt is a unit of power and J is a unit of Energy. 1 Watt is equal to $1Js^{-1}$. Therefore $1Gwh = 10^9 \times 60 \times 60 = 3.6 \times 10^{12}$J. This is a measure of energy.

Question 2d: C

The efficiency of the power stations is relevant to how much electricity can be generated so is not relevant to this difference. The efficiency of appliances is also not relevant to this difference as this applies after the energy is consumed. The difference between the amount of electricity generated and the amount consumed can be attributed to energy loss in distribution.

Question 3a: D

At $t = 0$ the velocity is zero. This velocity is positive for $t > 0$. The gradient is $\frac{dv}{dt} = a(t - T)^2 + 2at(t - T)$. At $t = 0$, this is equal to aT^2. The gradient of A at $t = 0$ is zero so this can be discounted. B and E can be discounted due to the negative velocity. C does not have the object at rest for $t > T$ so this is not the answer. D satisfies all the requirements and is the correct answer.

Question 3b: D

Use $\quad a = \frac{dv}{dt} = a(t - T)^2 + 2at(t - T) = a(t^2 - 2tT + T^2 + 2t^2 - 2tT) = a(3t^2 - 4tT + T^2) = a(3t - T)(t - T)$

Question 3c: E

Take the derivative of the acceleration. $\frac{da}{dt} = 3a(t - T) + a(3t - T) = 6at - 4aT = 2a(3t - 2T)$. Therefore, the maximum negative acceleration is at $t = \frac{2}{3}T$. Substitute this into the equation for the acceleration giving $\frac{dv}{dt} = aT\left(-\frac{1}{3}T\right) = -aT^2/3$

Question 3d: A

There is only one stationary point, so the maximum positive acceleration must occur at one of the boundaries of the domain i.e. either $t = 0$ or $t = T$. For $t = 0$, $\frac{dv}{dt} = aT^2$. At $t = T$, $\frac{dv}{dt} = 0$. Therefore, the answer is A.

Question 3e: C

Integrate the velocity distribution with respect to time to find the displacement. $v = a(t^3 - 2Tt^2 + T^2t)$. Integrating this gives $x = a\left(\frac{t^4}{4} - \frac{2Tt^3}{3} + \frac{T^2t^2}{2}\right)$. However, the object is at rest for $t > T$ so the displacement is the same at $t = 2T$ as it is at $t = T$. Inserting $t = T$ gives $x = a\left(\frac{T^4}{4} - \frac{2T^4}{3} + \frac{T^4}{2}\right) = aT^4(3 - 8 + 6)/12 = aT^4/12$.

Question 4a: D

The gravitational energy gained by the car is MgH. This is converted into kinetic energy of $\frac{1}{2}Mv^2$. Thus, $MgH = \frac{1}{2}Mv^2$. Rearranging this gives $v = \sqrt{2gH}$.

Question 4b: E

At the top of the loop the car has dropped a height of $H - 2R$ relative to the starting point. The loss in GPE is again converted into kinetic energy so $Mg(H - 2R) = \frac{1}{2}Mv^2$. This yields $v = \sqrt{2g(H - 2R)}$

Question 4c: B

If there were no acceleration the object would move in a straight line out of the circle. To maintain its circular trajectory the acceleration must be towards the centre of the circle.

Question 4d: D

For the car to remain in contact with the track at B the centripetal acceleration must be at least the gravitational acceleration. The velocity at B is given by $\sqrt{2g(H - 2R)}$. The acceleration is then $\frac{2g(H-2R)}{R} > g$. Rearranging this gives $2(H - 2R) > R$. This yields $H > \frac{5R}{2}$.

END OF PAPER

2016

Section 1

___

Question 1: G

Rearranging gives $\frac{x}{2} < 14$. Multiply by 2 to get $x < 28$.

Question 2: D

An alpha decay (emission of two protons and two neutrons) reduces the mass number by four and the proton number by two. Beta decay leaves the mass number unchanged and increases the proton number by one as a neutron is converted to a proton and an electron. Therefore, there has been 1 alpha decay and 2 beta decays.

Question 3: B

Expand $(\sqrt{3} - \sqrt{2})(\sqrt{3} - \sqrt{2}) = 3 - 2\sqrt{6} + 2 = 5 - 2\sqrt{6}$. This is the same as option B.

Question 4: F

The kinetic energy of an object is proportional to the square of the velocity so is not a straight line. The potential energy of an object being lifted increases linearly with height as $GPE = mgh = 20kg \times 10ms^{-2} \times 2m = 400J$. Therefore, the graph cannot represent option 2. If the resultant force is constant, the acceleration of the object is constant, so the velocity increases linearly with time. $a = \frac{F}{m} = 100/20 = 5ms^{-2}$. Work done is given by $W = Fd$. For a constant speed, d increases linearly with time. $W = 5N \times 2m = 10J$ so, the graph could represent number 4 as well.

Question 5: C

If $\frac{Q}{R} = \frac{5}{2}$ and $\frac{R}{S} = \frac{3}{10}$, then multiply the first equation by $\frac{R}{S}$, giving $\frac{Q}{S} = 15/20 = 3/4$. Therefore, $Q : S = 3 : 4$

Question 6: C

Require the total number of nucleons to be conserved. In the first diagram there are initially 236 nucleons: 235 in the U nucleus and 1 proton. After fission there are $144 + 89 + 2 = 235$ so nucleon number hasn't been conserved. In diagram 2 there are 236 nucleons initially and $137 + 96 + 3 = 236$ nucleons so this process is possible. In diagram 3, there are 236 at first and $145 + 87 + 3 = 235$ afterwards so this process can be discounted.

Question 7: E

If the mean age of 20 people is 28, the sum of all their ages is $20 \times 28 = 560$. The new mean age is 30 for 22 people, so the sum of the ages after the members join is $22 \times 30 = 660$. The two new members have added 100 years between them so the mean age of the two is 50 years

Question 8: D

The current through the circuit is given by $I = V/R = 24/(5 + R_V)$, where R_V is the resistance of the variable resistor. The power dissipated in the 5Ω resistor is $P = I^2R = \frac{24^2 \times 5}{(5+R_V)^2}$. To maximise the power P, we want to minimise the denominator and minimise R_V. Therefore, we choose $R_V = 3\Omega$ and insert this giving $P = \frac{24^2 \times 5}{8^2} = 3^2 \times 5 = 45W$

Question 9: C

If the scanner decreases by 20% each year its new value is 80% of its previous value. After two year its new value is $15000 \times 0.8 \times 0.8 = 9600$. Therefore, the value has decreased by $15000 - 9600 = £5400$.

Question 10: D

As $P = kR^2T^4$, an increase of R by a factor of 100 will increase P by a factor of $100^2 = 10^4$. The temperature decreasing by 50% will cause the power to become $\left(\frac{1}{2}\right)^4 = \frac{1}{16}$ of its previous value. Therefore, the new power will be $\frac{4.0}{16} \times 10^{26} \times 10^4 = 0.25 \times 10^{30} = 2.5 \times 10^{29}W$

Question 11: B

The interior angle at B is 30 degrees and the one at A is 60 degrees. This means the angle at C must be a right angle as $180 - 60 - 30 = 90$. Therefore, we can use $\sin\theta = \frac{O}{H} = \frac{BC}{AB} = \sin 60 = \sqrt{3}/2$. Rearranging this gives $BC = 2\sqrt{3}$km.

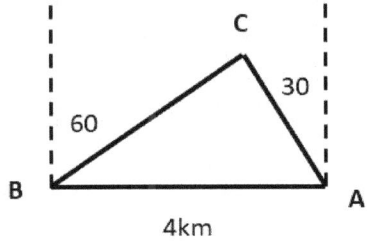

Question 12: E

The wave is transverse, so the particles of the medium oscillate perpendicular to the motion of the wave. In one minute there are $5 \times 60 = 300$ oscillations. The amplitude measures from the equilibrium position to a maximum. In one full oscillation a particle travels 4 times the amplitude i.e. 12cm. In one minute a particle of the medium will travel $300 \times 12 = 3600$cm

Question 13: F

For $x \propto \frac{1}{\sqrt{y}}$, $x = \frac{k}{\sqrt{y}}$. Insert $x = 8, y = 9$ giving $k = 24$. Rearranging the equation gives $\sqrt{y} = k/x = 4$. Therefore, $y = 16$.

Question 14: H

The tension is constant throughout the rope. Consider first the forces acting on the mass: $F = ma = 5 \times 0.8 = 4N = T - mg$. Therefore, the tension is $T = 54N$. The tension acts downwards on both sides of the pulley so the force on the coupling is $2T = 108N$.

Question 15: E

Split the trapezium into a rectangle and a triangle with areas of $x(x - 1)$ and $\frac{1}{2}x \times 6 = 3x$. i.e. $120 = x(x - 1) + 3x = x^2 + 2x$.

Rearranging gives $x^2 + 2x - 120 = 0$. The factors of 120 that differ by 2 are 12 and 10 so $(x + 12)(x - 10) = 0$. Discard the negative solution as negative lengths are not possible. Therefore $x = 10cm$ and RS has length 15cm.

Question 16: F

The current through the circuit is given by $I = V/R = 6/15 = 0.4A$. The power dissipated by the heater is $180/(3 \times 60) = 1W$. Now use $P = IV$ to find $V = 2.5V$. Current is defined as the charge per unit time so the charge passing through the heater in 3 minutes is $0.4 \times 180 = 72C$

Question 22: A

The distance travelled is given by the area under the velocity-time graph. In the first 20 seconds the object travels $\frac{1}{2} \times 8 \times 20 = 80$m. In the following 10 seconds the object travels $\frac{1}{2} \times 2 \times 10 = 10$m in the opposite direction. The total distance travelled is 90m and the object is 70m from its starting position after 30s. The average speed does not depend on the direction of travel as speed is a scalar quantity. Therefore, the average speed is $90/30 = 3$ms^{-1}

Question 23: D

There are 120 students: 46 girls and 74 boys. $\frac{2}{3} \times 36 = 24$ boys chose tennis and 12 girls chose tennis. 25 girls chose to swim and the $46 - 25 - 12 = 9$ remaining girls did archery. 27 students did archery- 9 girls and 18 boys. The remaining 32 boys all chose swimming. Therefore, if a boy is chosen at random the chance he swims is $\frac{32}{74} = \frac{16}{37}$.

Question 24: G

The mass of each component is given by $m = \rho V$, so the total mass per unit volume is $0.9X + 0.1Y$. The fraction of the mass that is tin is $\frac{0.1Y}{0.9X+0.1Y} = \frac{Y}{9X+Y}$. Therefore, the percentage mass of tin is $\frac{Y}{9X+Y} \times 100$

Question 25: D

$$\frac{9^{2n+1} \times 3^{4-3n}}{27^{2-n}} = 3^{4n+2} \times 3^{4-3n} \times 3^{3n-6}$$

This equals $3^{6+n+3n-6} = 3^{4n}$

Question 26: D

The initial and final momentum must be the same i.e. 0. The mass of the alpha particle is $\frac{4}{234}$ that of the thorium. Therefore $v_\alpha = \frac{234}{4} v_{Th}$. The kinetic energy of the Thorium is given by $E_{Th} = \frac{1}{2} m_{Th} v_{Th}^2$. The kinetic energy of the alpha particle is $E_\alpha = \frac{1}{2} \times \frac{4m_{Th}}{234} \times \left(\frac{234 v_{Th}}{4}\right)^2 = \frac{234}{4} E_{Th}$. Summing these gives $E = E_\alpha + \frac{4}{234} E_\alpha$

This yields $E_\alpha = \frac{234E}{238}$.

Question 27: E

The external angle of a regular polygon with n sides is $\frac{360}{n}$. This corresponds to the angle RQT. As the triangle is isosceles the angle RTQ is the same as the angle RQT. Therefore, the angle $x = 180 - \frac{2 \times 360}{n}$. Rearranging this gives $\frac{720}{n} = 180 - x$. Further manipulation yields $n = \frac{720}{180-x}$

Question 28: G

For the sound to reflect from the left building and back to the student the wave must travel 96m. This takes 0.3s. For the sound wave to reflect from the right building the sound wave would have to travel 160m, taking 0.5s. The first click occurs at 0s then another must occur 0.3s later and 0.2s after that. The lowest frequency that would produce clicks with both of these intervals is 10Hz i.e. 1 click every 0.1s.

Question 29: C

Sub in $x = 2$: $8 + 4p + 2q + p^2 = 0$

With $x = 1$: $1 + p + q + p^2 = -3.5$

Multiply the second equation by 2 and subtract it from the first

$$6 + 2p - p^2 = 7$$

$$p^2 - 2p + 1 = 0$$

$$(p - 1)^2 = 0$$

Therefore the only solution is $p = 1$

Question 30: C

Newton's third law: For every action there is an equal and opposite reaction. This only applies to the internal forces in this diagram as the particles of the air have not been considered. The only internal forces are N and P which are equal and opposite. The answer is 4 only.

Question 31: C

The square has a side length 6. The centre of the square is $(-2, 3)$. The diameter of the circle is also 6 so it has a radius 3. The equation for a circle radius, r, and centred at (x_0, y_0) has the equation $(x - x_0)^2 + (y - y_0)^2 = r^2$. Inserting the values gives

$$(x + 2)^2 + (y - 3)^2 = 9$$

$$x^2 + 4x + y^2 - 6y + 4 = 0$$

Question 32: C

The forces acting on the mass are its weight downwards and the tension upwards: $F = T - mg = ma = 1600N$. Rearranging gives $T = 1600 + 8000 = 9600N$

Question 33: C

For a series $a_1 = 8$. $a_5 = 2$. A geometric series means consecutive terms differ by the same factor. $8 \times r^4 = 2$. $r = \pm (1/4)^{1/4} = \pm \frac{1}{\sqrt{2}}$. As the sixth term is real and positive we can discard the imaginary and negative roots. So $r = \frac{1}{\sqrt{2}}$. Using this gives the series sum as $\frac{8}{1-\frac{1}{\sqrt{2}}} =$

$$\frac{8\sqrt{2}}{\sqrt{2}-1} = \frac{8\sqrt{2}(\sqrt{2}+1)}{2-1} = 8(2 + \sqrt{2})$$

Question 34: A

The vertical component of the force applied to the trolley is $50 \sin 37 = 50 \times 0.6 = 30N$. The horizontal component is $50 \cos 37 = 40N$. The total force downwards is the sum of the vertical component of the force applied by the shopper and the trolley's weight i.e. $30 + 350 = 380N$. This is balanced by the reaction force from the surface. The work done by the pushing force is $W = Fd = 40N \times 15m = 600J$

Question 35: A

Tangents to a circle are at 90 degrees to the radius of the circle at that point. Thereby we can form two right angled triangles with points at P, C and where the two tangents touch the circle. The hypotenuse of these triangles is 20cm long and the radii are 10cm. Using Pythagoras, the distance from P to the point on the circle is $\sqrt{20^2 - 10^2} = 10\sqrt{3}$cm. Now consider the angle formed by the line PC and from C to the tangent line: $\cos\theta = \frac{10}{20} = 0.5$. Therefore, $\theta = 60°$. The total area of the two right angled triangles is $2 \times \frac{1}{2} \times 10\sqrt{3} \times 10 = 100\sqrt{3}cm^2$. The removed area is $\frac{1}{3}$ of the area of the circle i.e. $\frac{100\pi}{3}$. Subtracting this area gives $\frac{100}{3}(3\sqrt{3} - \pi)$

Question 36: H

The moments of each object are given by $F \times d$. The total moment must be zero for balance. The resultant moment of the boy and the woman is $g(1.2 \times 35 - 0.8 \times 60) = -60$Nm with positive moments causing a clockwise rotation. The moment of the plank must be 60Nm to balance this so $d = \frac{60\text{Nm}}{15g} = 0.4$m to the right of the pivot. The total force between the pivot and the plank is the sum of the weights i.e. $F = (35 + 15 + 60)g = 1100$N

Question 37: D

Use $\tan \theta = \frac{\cos \theta}{\sin \theta}$ and multiply through by $\cos \theta$. This gives $7 \cos^2 \theta - 3 \sin^2 \theta = \cos \theta$. Now replace $\sin^2 \theta = 1 - \cos^2 \theta$ giving $10 \cos^2 \theta - \cos \theta - 3 = 0$. Substitute $x = \cos \theta$ and factorise to find $(5x - 3)(2x + 1) = 0$. The two solutions are $\cos \theta = \frac{3}{5}$ and $\cos \theta = -\frac{1}{2}$

Question 38: B

The initial energy is $E = \frac{1}{2}mv^2 + mgh = 100 \times 25 + 16000 = 18500J$. At point Y the new energy is $E_Y = 100 \times 81 + 4000 = 12100J$. The difference between these two must be the work done against resistive forces. This is 6400J

Question 39: G

Rearrange this equation to get $3x^2 - (a + 2)x + 3 = 0$. To find the number of distinct real roots the discriminant is required: $b^2 - 4ac$. For two real roots this must be greater than zero. $(a + 2)^2 - 36 > 0$. This gives $(a + 2)^2 > 36$. This happens when $a + 2 > 6$ or when $a + 2 < -6$. Therefore, we require $a > 4$ or $a < -8$

Question 40: B

The first must be untrue. The weight of the block acts vertically downwards so has no horizontal component. For the block to be stationary the forces must balance in all directions so frictional forces are required to balance the horizontal force P. The second is true as this allows the frictional force to balance P horizontally. The third is not true as the horizontal forces could not balance. The moment of force P around the edge in contact is $F \times l$ where l is the perpendicular distance. The distance between P and the table surface is $\frac{1}{2}(1 + \sqrt{3})d$ so the moment is greater than $P \times d$

Question 41: D

A line perpendicular to one of gradient m has a gradient $\frac{-1}{m}$. At $y = 0$, the first line satisfies $x = \frac{-3}{m}$. The second line crosses the x axis at $x = \frac{-2}{p} = 2m$. Therefore, $5 = 2m + \frac{3}{m}$ which gives $2m^2 - 5m + 3 = 0$. Factorising this gives $(2m - 3)(m - 1) = 0$. Two possible solution $m = 1$ and $m = \frac{3}{2}$. Discard the first solution as $m > 1$. This gives $p = -2/3$. $m + p = \frac{3}{2} - \frac{2}{3} = \frac{5}{6}$

Question 42: A

The gravitational potential energy is given by mgh with h the height of the centre of mass of the block. The first arrangement has GPE $= mg\left(\frac{b}{2} + \frac{3b}{2}\right) = 2mgb$. In the second arrangement GPE $= mg\left(\frac{a}{2} + \frac{3a}{2}\right) = 2mga$. Therefore, the expression is $2mg(a - b)$.

Question 43: G

f(x) is an increasing function when its derivative with respect to x is greater than zero. $\frac{df}{dx} = 3x^2 - a^2 > 0$. Rearranging gives $x^2 > \frac{a^2}{3}$. This is true when $x < \frac{a}{\sqrt{3}}$ or when $x < -\frac{a}{\sqrt{3}}$.

Question 44: G

Initially the object has kinetic energy $\frac{1}{2}m \times 8^2 = 32m$. When it has $v = 2ms^{-1}$ the kinetic energy is $\frac{1}{2}m \times 2^2 = 2m$. Therefore, the object has gained $30m = mgh$ of gravitational potential energy. This gives $h = 3$. Use $s = ut + \frac{1}{2}at^2 = 8t - \frac{1}{2}gt^2$. Insert $s = 3m$ gives $5t^2 - 8t + 3 = 0$. Factorising $(5t - 3)(t - 1) = 0$. The two possible solutions are $t = 0.6s$ and $t = 1s$. The smaller value of t is as the object is on the way up and the larger value as the object is falling. Therefore, G is the correct answer.

Question 45: E

Translation by the vector (4,3) gives the new equation $y - 3 = (x - 4)^2$.

To reflect in $y = -1$, isolate $y + 1$ on the LHS i.e. $y + 1 = 4 + (x - 4)^2$. A reflection in $y = -1$, causes the sign of the RHS to be changed giving $y + 1 = -4 - (x - 4)^2$. Rearranging this yields $y = -5 - (x - 4)^2$.

Question 46: E

Momentum must be conserved in the collision. The initial momentum is $4 \times 10 = 40 \text{kgms}^{-1}$. If P has velocity v after the collision the final momentum is $4v + 20$. Therefore, $4v = 20$. $v = 5 \text{ms}^{-1}$. This is to the right. The initial energy is $\frac{1}{2} \times 4 \times 100 = 200 \text{J}$. The kinetic energy after the collision is $\frac{1}{2} \times 2 \times 100 + \frac{1}{2} \times 4 \times 25 = 100 + 50 = 150 \text{J}$. The energy lost is 50J.

Question 47: C

$$2x^4 - 9x^2 + 4 > 0$$

Substitute $y = x^2$, gives $2y^2 - 9y + 4 > 0$. Factorise this to get $(2y - 1)(y - 4) > 0$. Two roots are at $y = \frac{1}{2}$ and $y = 4$. For y large positive or negative the quadratic will be greater than zero. Therefore, $x^2 < \frac{1}{2}$ or $x^2 > 4$ satisifies the original inequality. For $x^2 < \frac{1}{2}$, require $-\frac{1}{\sqrt{2}} < x < \frac{1}{\sqrt{2}}$. For $x^2 > 4$, require $x < -2$ or $x > 2$.

Question 48: A

Add the two vectors together at 90 degrees to each other. The magnitude of the resultant force is $\sqrt{9^2 + 12^2} = 15N$. The maximum friction is $F = \mu R = 0.25 \times 20 = 5N$. As the object moves the net force is $F = 10N = 2 \times a$. This gives $a = 5ms^{-2}$.

Question 49: B

The distance parallel to the y axis between the lines $y = 10$ and $y = 4x^3 - 12x^2 - 36x - 15$ is $f(x) = 25 - 4x^3 + 12x^2 + 36x$. Want to find the minimum of $f(x)$ for negative values of x.

$$\frac{df}{dx} = -12x^2 + 24x + 36 = -12(x^2 - 2x - 3) = -12(x - 3)(x + 1)$$

Two stationary points: $x = 3$ and $x = -1$. The highest point of the cubic y for negative x is at $x = -1$, where $f = 25 + 4 + 12 - 36 = 5$

Question 50: C

Consider first the 30kg mass. This is accelerating at 2.5ms^{-2} so the net force is $F = 75\text{N} = mg - T = 300 - T$. This gives $T = 225\text{N}$. Now consider the 20kg mass. The forces acting parallel to the plane are the tension up the plane, $mg \sin \theta$ down the plane and friction down the plane. $225 - 100 - F_{\text{Friction}} = ma = 50$. Rearranging gives $F_{\text{Friction}} = 75\text{N}$.

Question 51: A

Expand $x = 3(y-1)^2 + 4 = 3(y^2 - 2y + 1) + 4 = 3y^2 - 6y + 7$. The two lines intersect at $7 = 3y^2 - 6y + 7$ i.e. $3y^2 - 6y = 3y(y-2) = 0$. Integrate from $y = 0$ to $y = 2$ the difference $f(y) = 3y^2 - 6y$. $\int_0^2 dy(3y^2 - 6y) = [y^3 - 3y^2]_0^2 = 8 - 12 = -4$. Only need the magnitude of the area so discard the negative.

Question 52: E

Initially the momentum is 4000×7425. Call the mass of ejected fuel m. After the fuel is ejected momentum conservation requires

$$4000 \times 7425 = 7500(4000 - m) + 6000m$$

$$4 \times 7425 = 7.5(4000 - m) + 6m$$

$$4 \times 7425 = 30000 - 1.5m$$

$$1.5m = 30000 - 29700 = 300$$

$$m = 200kg$$

Question 53: C

To find the number of roots of the equation consider the locations of the turning points. $\frac{dy}{dx} = 12x^3 - 12x^2 - 24x = 12x(x^2 - x - 2) = 12x(x - 2)(x + 1)$

Therefore, stationary points are at $x = -1, 0, 2$.

Define $f = 3x^4 - 4x^3 - 12x^2 + 20 - k$.

$f(-1) = 3 + 4 - 12 + 20 - k = 15 - k$

$f(0) = 20 - k$

$f(2) = 48 - 32 - 48 + 20 - k = -12 - k$

For 4 distinct roots we require stationary points to alternate above and below the x axis. Need $15 - k < 0$, $20 - k > 0$ and $-12 - k < 0$. i.e. $k > 15$, $k < 20$ and $k > -12$. The required conditions are $15 < k < 20$.

Question 54: B

For the object to be stationary the forces must cancel in all directions. The string must be at an angle such that the horizontal component of the tension is 30N. Call the angle the string makes with the vertical θ. The vertical component of the tension must balance the weight of the object i.e. 40N. The magnitude of the tension in the string is $\sqrt{30^2 + 40^2} =$ 50N. This gives $\sin\theta = 0.6$ and $\cos\theta = 0.8$. The vertical distance from the ceiling to the block is now $0.35\cos\theta = 0.28$m. Therefore, the change in height is 0.07m. The change in GPE is $mg\Delta h = 40 \times 0.07 = 2.8$J

Section 2

Question 1a: A

The relation between the tension in the wire and its extension is $F = kx$ where k is the stiffness. This means that the largest value of $k = \frac{F}{x}$ will correspond to the largest gradient on the graph i.e. S1.

Question 1b: B

The strain of a wire is defined as $\epsilon = \frac{x}{L}$ where x is the extension of the wire and L is its natural length. The samples have natural length 100mm so at 2% strain then the extension is 2mm. Sample that obey Hooke's Law have a Tension that is directly proportional to the extension giving a straight line through the origin. Samples S1, S3, S4 and S5 obey this well beyond 2mm so the answer is S2

Question 1c: A

Young's Modulus is $E = \frac{\sigma}{\epsilon} = \frac{FL}{xA}$. The ratio $\frac{F}{x}$ is the stiffness which is the gradient of the graph $k = \frac{250}{5 \times 10^{-3}} = 5 \times 10^4 \text{Nm}^{-1}$. $\frac{L}{A} = \frac{100 \times 10^{-3}}{(5 \times 10^{-3})^2} = 4 \times 10^3 \text{m}^{-1}$. Multiplying these gives $E = 20 \times 10^7 \text{Pa} = 200 \text{MPa}$.

Question 1d: A

The work done against the tension T is $W = \int dx\, T$. This gives $W = \frac{ax^2}{2} - \frac{bx^3}{3} + c$. Set $c = 0$ as no work has been done when the extension is zero. At $x = 10mm = 10^{-2}m$ the work done is $W = \frac{a}{2} \times 10^{-4} - \frac{b}{3} \times 10^{-6} Nm$

Question 2a: F

Currents are equal for components that are in series whereas voltages are equal across components that are in parallel. Therefore, the voltages across R_2 and R_3 are the same.

Question 2b: A

To find the current we need to find the total resistance of the circuit. First consider the two resistors in parallel. The total resistance of the resistors in parallel is $\frac{1}{R_T} = \frac{1}{R_2} + \frac{1}{R_3}$. This yields $R_T = \frac{R_2 R_3}{R_2 + R_3}$. The total resistance of the circuit is then the sum of the two resistances in series: $R_{Tot} = R_1 + R_T = \frac{R_1(R_2 + R_3) + R_2 R_3}{R_2 + R_3} = \frac{R_1 R_2 + R_1 R_3 + R_2 R_3}{R_2 + R_3}$. This gives the current through the voltage source as $I = \frac{V}{R_{Tot}} = \frac{V(R_2 + R_3)}{R_1 R_2 + R_1 R_3 + R_2 R_3}$

Question 2c: D

We know the current through the first resistor from the previous part which is now $I = \frac{V(R_1+R_3)}{R_1^2+2R_1R_3}$. The voltage across the first resistor is then $V_1 = IR_1 = \frac{V(R_1+R_3)}{R_1+2R_3}$. Subtracting this from the total voltage gives the voltage across each of the resistors in parallel: $V_3 = \frac{V(R_1+2R_3-R_1-R_3)}{R_1+2R_3} = \frac{VR_3}{R_1+2R_3}$. The power dissipated across R_3 is then $P = \frac{V_3^2}{R_3} = \frac{V^2R_3}{(R_1+2R_3)^2}$

Question 2d: B

Consider $\frac{1}{P} = \frac{(R_1+2R_3)^2}{V^2R_3}$ and take the derivative with respect to R_3. $\frac{d}{dR_3}\left(\frac{1}{P}\right) = \frac{2\times2\times R_3(R_1+2R_3)-(R_1+2R_3)^2}{R_3^2} = \frac{R_1+2R_3}{R_3^2} \times (4R_3 - R_1 - 2R_3)$. The derivative is set to zero for a stationary point which only occurs when $(4R_3 - R_1 - 2R_3) = (2R_3 - R_1) = 0$. This is when $R_3 = \frac{1}{2}R_1$.

Question 3a: C

$1\text{ns} = 10^{-9}\text{s}$. The speed of light is $c = 3 \times 10^8 \text{ms}^{-1}$. This gives a distance of $3 \times 10^8 \times 10^{-9} = 0.3\text{m}$

Question 3b: F

The speed of light in a material with refractive index n is $v = \frac{c}{n}$. For n = 1.5, $v = 2 \times 10^8 \text{ms}^{-1}$. Therefore, the time taken to travel 9km is $T = \frac{9 \times 10^3}{2 \times 10^8} = 4.5 \times 10^{-5}\text{s} = 45\mu\text{s}$.

Question 3c: A

A larger refractive index leads to a smaller speed of propagation, so the speeds obey the inequality $v_{nom} < v_{blue} < v_{red}$. The faster the light travels the shorter the propagation time will be, so the answer is $T_{red} < T_{blue} < T_{nom}$.

Question 4a: E

At the point B the cyclist has kinetic energy $E + Mgh$ sourced from the work done by the cyclist and from gravitational potential energy. This gives the cyclist a speed of v where $\frac{1}{2}Mv^2 = E + Mgh$. This gives $v = \sqrt{\frac{2E}{M} + 2gh}$. The component down the slope is $\sqrt{2\left(\frac{E}{M} + gh\right)} \times \cos\theta$.

Question 4b: B

Consider the motion of the cyclist perpendicular to the plane. The initial velocity in this direction is $V \sin \theta$ and the component of the gravitational acceleration is $g \cos \theta$. Now use $s = ut + \frac{1}{2}at^2$ and set $s = 0$ as this is when the bike is in contact with the plane. This gives $0 = Vt \sin \theta - \frac{1}{2}gt^2 \cos \theta$. Factorise this to get $0 = t\left(V \sin \theta - \frac{gt}{2} \cos \theta\right)$. The solution $t = 0$ corresponds to when the cyclist leaves the ground. The solution $t = \frac{2V}{g} \tan \theta$ is when the cyclist lands at C.

Question 4c: C

Consider the components parallel to the slope. There is an acceleration of $g \sin \theta$ down the slope and an initial velocity of $V \cos \theta$. Insert these values and the time from the previous part into $s = ut + \frac{1}{2}at^2$ giving

$$L = V \cos \theta \times \frac{2V}{g} \tan \theta + \frac{1}{2} g \sin \theta \left(\frac{2V}{g} \tan \theta\right)^2 = \frac{2V^2 \sin \theta}{g}(1 + \tan^2 \theta).$$

Use $1 + \tan^2 \theta = 1/\cos^2 \theta$. This gives $L = \frac{2V^2 \sin \theta}{g \cos^2 \theta}$

Question 4d: D

The maximum perpendicular distance of the cyclist from the plane occurs at the time halfway between it leaving the plane and landing i.e. $t = \dfrac{V}{g}\tan\theta$. Insert this into $s = Vt\cos\theta + \dfrac{1}{2}gt^2\sin\theta$. This gives $s =$

$$V\cos\theta \times \frac{V}{g}\tan\theta + \frac{1}{2}g\sin\theta \times \left(\frac{V}{g}\tan\theta\right)^2 = \frac{V^2}{g}\sin\theta +$$

$$\frac{V^2}{2g}\sin\theta \tan^2\theta = \frac{V^2}{g}\sin\theta\left(1 + \frac{1}{2}\tan^2\theta\right).$$

END OF PAPER

2017

Section 1

Question 1: F

$\sqrt{12} = 2\sqrt{3}$. Inserting this into the expression gives $\frac{\left(3\sqrt{3}\right)^2}{\left(\sqrt{3}\right)^2} = 3^2 = 9$

Question 2: B

The car decelerates between $t = 110s$ and $t = 130s$. The distance travelled is the area under the graph between these times. This is equal to $20 \times 20 + \frac{1}{2} \times 20 \times 10 = 500m$.

Question 3: E

Move all terms to the LHS to get $2x^2 + x - 15 \geq 0$. Factorise this equation to get $(2x - 5)(x + 3) \geq 0$. The equation $2x^2 + x - 15 = 0$ has roots at $x = 2.5$ and $x = -3$. When x is a large positive or negative number x^2 dominates so the inequality is satisfied when x tends to $\pm\infty$. Therefore, $x \leq -3$ or $x \geq 2.5$ solves the inequality.

Question 4: G

When the water is heated its density decreases as its volume increases. This means that the mass in a fixed volume decreases. Statements 2 and 3 are true.

Question 5: B

Add 5 to both sides and divide by 3: $\frac{y+5}{3} = \left(\frac{x}{2} - 1\right)^2$. Square roots give $\pm\sqrt{\frac{y+5}{3}} = \frac{x}{2} - 1$. Add 1 and multiply by 2 to get $x = 2 \pm 2\sqrt{\frac{y+5}{3}}$

Question 6: D

The car gains potential energy $mgh = 1200 \times 10 \times 1 = 12kJ$. All the other energy inputted is lost to the surroundings so the total energy lost is $28 - 12 = 16kJ$

Question 7: G

Sam gives $2x + 5y = P$. Lesley: $3x + 2y = Q$. Multiply the first equation by 3 and the second by 2: $6x + 15y = 3P$ and $6x + 4y = 2Q$. Subtract the second equation from the first to eliminate x: $11y = 3P - 2Q$. Rearranging this gives $y = \frac{3P-2Q}{11}$

Question 8: D

The time difference corresponds to a difference in distance travelled of $d = 3.0 \times 10^8 \times 4.0 \times 10^{-10} = 1.2 \times 10^{-1}$m. If the source is a distance x to the left of Q the gamma ray must travel $3 - x$ to the left detector and a distance $3 + x$ to the right detector. So, the source is $\frac{d}{2} = $ 6cm away from the midpoint Q.

Question 9: E

$P = kQ^2$. Insert $P = 2$, $Q = 4$ to get $P = Q^2/8$. Secondly, $Q = \frac{k_2}{R}$. Insert $Q = 2$, $R = 5$ to find $k_2 = 10$. Substitute $Q = \frac{10}{R}$ into the equation for P to get

$$P = \frac{100}{8R^2} = \frac{25}{2R^2}.$$

Question 10: B

In fission the total number of nucleons and the number of protons must be conserved. These conditions give equations $240 = w + y + z$ and $94 = 54 + x$. Rearranging the first equation gives $z = 240 - (w + y)$

Question 11: D

Factorising the numerator gives $2 - \frac{(3x-2)(3x+2)}{x(2-3x)} = 2 + \frac{3x+2}{x} = 5 + \frac{2}{x}$

Question 12: F

The power supplied is $P = IV$. The work done lifting the mass is $mgh = 20 \times 10 \times 6 = 1200J$. If the system is 80% efficient then the total input energy is $\frac{1200}{0.8} = 1500J$. The power is $P = \frac{W}{\Delta t} = \frac{1500}{5} = 300W$. The current is therefore $I = \frac{300}{12} = 25A$

Question 13: D

Rewrite the expression using $4 = 2^2, 8 = 2^3$ and $\sqrt{2} = 2^{1/2}$,

$2^{3+2x}2^{2x}2^{-3x} = 2^{5/2}$. Take logarithms to find $3 + 2x + 2x - 3x = 5/2$.

$x = -0.5$.

Question 14: A

Emission of an alpha particle would cause the mass number to decrease by 4 and the proton number to decrease by 2. Emission of a beta particle leaves the mass number unchanged by increases the proton number by 1. If the alpha particle were emitted first the mass number would be P − 4 and proton number Q − 2. Emission of 3 beta particles would then increase the atomic number to Q − 1, Q and Q+1. If the beta particles were emitted first the atomic mass would be P and the atomic number would increase to Q + 1, Q + 2, Q + 3. The only one that does not feature is A.

Question 15: F

There are 3X girls and 100 students in total. The number of girls studying Spanish is 35 − Y. The number of girls studying German is 3X − X − (35 − Y) = 2X + Y − 35 i.e. Total number of girls minus girls studying French and Spanish. Adding the number of boys doing German gives 2X + 3Y − 35.

Question 16: A

Density is given by $\rho = M/V$, where the volume $V \propto r^3$. Therefore, $\rho \propto r^{-3}$. The mass can be assumed to be constant as it is dominated by the nucleus of the atom. As the radius of the iron atom is 3×10^4 times that of the nucleus, the density of the atom is $(3.0 \times 10^4)^{-3}$ times that of the nucleus.

Question 17: C

The exterior angle of an n-sided polygon is $\frac{360}{n}$.

Therefore, we can write $\frac{360}{n} - \frac{360}{n+3} = 4$. Multiply by $n(n + 3)$ to get:

$360(n + 3) - 360n = 4n(n + 3)$.

Divide by 4 and expand to get: $n^2 + 3n - 270 = 0$. Factorise to find $(n - 15)(n + 18) = 0$. This gives $n = 15$ or $n = -18$. Can discard the negative value so the answer is C.

Question 18: E

The wave equation is $v = f\lambda$. Between time t_1 and t_2 there is $3/2$ full waves so the period is $\frac{2}{3}(t_2 - t_1)$ and the frequency is $\frac{3}{2(t_2 - t_1)}$. The distance between x_2 and x_1 is half a wavelength. Combining these gives $v = \frac{3(x_2 - x_1)}{t_2 - t_1}$.

Question 19: B

The angle SLR is 40 degrees. The angle LRC is the same as SLR so is 40 degrees. The lengths CL and RC are the same, so the triangle is isosceles so the angles LRC and CLR are the same. These are both 40 degrees so the angle LCR is $180 - 80 = 100$. The bearing of L from C is given by $180 - 100 = 80$ degrees.

Question 20: C

The power is $P = IV$. This gives a current $I = 150/12 = 50/4 = 12.5A$. Current is defined as charge per unit time so the charge passing through the element is $12.5 \times 20 \times 60 = 250 \times 60 = 15000C$.

Question 21: B

At 4:40 the minute hand points towards 8 and the hour hand has moved $\frac{2}{3}$ of the way towards 5. Each hour subtends an angle of 30 degrees. Between 8 and 5 there are 90 degrees. Between 5 and the hour hand is $\frac{1}{3} \times 30 = 10$. Therefore, the angle between the two hands is $90 + 10 = 100$ degrees.

Question 22: C

Momentum must be conserved in the collision. $M_f \times 2 = M_P \times 5$, where M_f is the mass of the freight train and M_P the mass of the passenger train. The mass of the passenger train is $M_P = \frac{2M_f}{5}$ and $M_f = 390 + 210 = 600$. This gives $M_P = 240$. Therefore, $240 - 140 = 100$ tonnes of passenger carriages. This is 10 carriages.

Question 23: C

There are $4 + x$ rabbits. The chance that the first rabbit is male is $\frac{x}{x+4}$. Now that one male rabbit has been removed the chance the second rabbit is male is $\frac{x-1}{x+3}$. This gives $\frac{x(x-1)}{(x+4)(x+3)} = \frac{1}{3}$. Rearranging this gives, $3x(x-1) = (x+4)(x+3)$.

$$3x^2 - 3x = x^2 + 7x + 12$$

$$2x^2 - 10x - 12 = 0$$

Factorising gives $2(x-6)(x+1) = 0$. Therefore $x = 6$ or $x = -1$. Take the positive value.

Question 24: C

The parachutist has kinetic energy $\frac{1}{2}mv^2 = 72 \times 5^2/2 = 900J$. The gravitational potential energy lost is $mg\Delta h$. In one second the parachutist loses 5m of height so the rate of change of GPE is $72 \times 10 \times 5 = 3600Js^{-1}$. The air resistance's third law pair is the force that acts downwards on the air particles not the gravitational force.

Question 25: F

First calculate the radius of the circle by considering the distance from the point O to where the corner of the square and the circle meet. This distance is $\sqrt{x^2 + \left(\frac{x}{2}\right)^2} = \frac{x\sqrt{5}}{2}$. The shaded area is half the area of the circle with the area of the square subtracted i.e. $\frac{5\pi x^2}{8} - x^2 = \left(\frac{5\pi - 8}{8}\right) x^2$

Question 26: D

In six hours X has undergone 2 half-lives and Y has undergone 3 half-lives. Originally X was $\frac{1}{2}$ of the mixture but it is now $\frac{1}{2} \times \frac{1}{4} = \frac{1}{8}$. Similarly, Y now makes up $\frac{1}{16}$ of the mixture. The remaining is Z i.e. $\frac{13}{16}$.

Question 27: G

The volume of a cylinder is $\pi r^2 h$. The volume of the pipe is $16\pi(5^2 - 4^2) = 16 \times 9\pi \, cm^3$. Multiply this by the density to get $1152\pi g$.

Question 28: E

Kinetic energy is given by $\frac{1}{2}mv^2$. The mass of car Y is M and its velocity is v. Y has kinetic energy $\frac{1}{2}Mv^2$. Car X has energy $\frac{1}{2} \times \frac{4}{5}M \times \left(\frac{3}{2}v\right)^2 = \frac{1}{2}Mv^2 \times \frac{9}{5}$. Therefore, X has 1.8 times the kinetic energy of car Y.

Question 29: E

$$1 - \left(\frac{3+\sqrt{3}}{6-2\sqrt{3}}\right)^2 = 1 - \frac{9+3+6\sqrt{3}}{36+12-24\sqrt{3}} = 1 - \frac{2+\sqrt{3}}{4(2-\sqrt{3})} = 1 - \frac{(2+\sqrt{3})^2}{4(2-\sqrt{3})(2+\sqrt{3})} =$$

$$1 - \frac{7+4\sqrt{3}}{4} = \frac{-3}{4} - \sqrt{3}.$$

Question 30: B

The moment of a force about the point P is given by $F \times d$, with d the perpendicular distance. If the load is moved to the left by 5m then the moment around P has increased by $4000 \times 5 = 20000$Nm. To balance this the counterweight must move such that its moment increases by the same amount but in the opposite direction. Thus, it must move to the right by $\frac{20000}{2000 \times 10} = 1$m

Question 31: C

The first equation can be rearranged to give $\sin x = -\frac{1}{2}$. Similarly, the second equation gives $\cos 2x = \frac{1}{2}$. $\cos 2x = 1 - 2\sin^2 x$ so, this can be simplified to $\sin^2 x = \frac{1}{4}$ i.e. $\sin x = \pm\frac{1}{2}$. Consider the graph of $\sin x$. There are two points where $\sin x = \frac{1}{2}$ for values of x smaller than that where $\sin x = -\frac{1}{2}$. Therefore, including the point at k there are 3 values of x that satisfy at least one of these equations.

Question 32: A

Initially the velocity is positive and then decreases as kinetic energy is converted into gravitational potential energy. When it reaches the top of it motion the velocity is zero and then the velocity will become negative. When the ball is caught the velocity should have the same magnitude as when it was thrown as no energy has been lost (neglecting air resistance). The acceleration of the ball is constant, so the velocity will change linearly with time, so the graph should be a straight line.

Question 33: B

Rewrite the equation $3(3^x)^2 - 6(3^x) = 0$. Then divide by 3 and substitute $y = 3^x$ to get $y^2 - 2y = 0$. This has solutions $y(y - 2) = 0$. Therefore, the two possible solutions are $3^x = 2$ and $3^x = 0$. The second requires $x = -\infty$ so can be ignored. Now take the logarithm with base 3 to get $x = \log_3 2$

Question 34: C

If the aircraft is travelling at a constant speed, then the resultant force on the aircraft must be zero.

Question 35: C

First find the radius of the circle. If the angle QOR is $\pi/6$ then the remaining angle is $2\pi - \pi/6 = 11\pi/6$. The long arc has length $11\pi r/6 = 22\pi$. This gives $r = 12$mm. Therefore, the distance SO and OP are both 30mm. The area of the triangle is $\frac{1}{2} ab \sin\theta = 450 \times \frac{1}{2} = 225$mm^2. The remaining area of the circle is $\frac{11}{12} \pi r^2 = 132\pi$. Adding these contributions gives $132\pi + 225$.

Question 36: E

The moments around the pivot must balance for the bar to be in equilibrium. It is subject to 2 moments: one arising from its weight and other from F. The moment due to its weight is $Mg \times d = 60 \times 10 \times 2 = 1200Nm$. The moment due to F is only due to the vertical component of F i.e. $F \sin 60$. The moment is $F \sin 60 \times 4$. Equating the two gives $F = \dfrac{300}{\sin 60}$

Question 37: F

The equations given can be rewritten $y = 2^{3p}$ and $z = 2^{-2q}$. Insert these into $\dfrac{y^3}{z^2} = 2^{9p+4q}$. Taking the logarithm base 2 of this gives $\log_2 \dfrac{y^3}{z^2} = 9p + 4q$

Question 38: D

Use $v^2 = u^2 + 2as$ with $u = 40, a = -14.4, s = 20$. This gives $v^2 = 1600 - 4 \times 144 = 1024$. Taking the square root gives $v = 32ms^{-1}$. It is easiest to do this by noticing $1024 = 2^{10}$.

Question 39: D

Take the derivative to find $\frac{dy}{dx} = 3x^2 + 2px + q$. At stationary points this equals zero. Inserting the two values of x gives $0 = 12 + 4p + q$ and $0 = 48 + 8p + q$. Subtracting the first equation from the second gives $4p + 36 = 0$ so $p = -9$. Using this in either of the first two equations gives $q = 24$.

Question 40: F

The force meter will measure the tension in the string. The tension is constant throughout as the system is in equilibrium. Consider the hanging mass. Here the tension must balance the weight of the object i.e. $T = 10 \times 1 = 10N$. This will be the reading on the meter.

Question 41: D

The area of the triangle in given by $\frac{1}{2}PQ\timesQR \times \sin 60 = 2x(8 - 3x) \times \frac{\sqrt{3}}{2} = \sqrt{3}x(8 - 3x)$. Take the derivative of this with respect to x and set this equal to zero giving $\sqrt{3}(8 - 6x) = 0$. This gives a maximum at $x = 4/3$. Substitute this into the equation for the area to get $A = \sqrt{3} \times \frac{4}{3} \times 4 = \frac{16\sqrt{3}}{3}$.

Question 42: A

As the apple falls it converts GPE into kinetic energy. However, some energy is dissipated by doing work against resistive forces. The initial GPE is mgh $= 0.1 \times 10 \times 4 = 4J$. When the apple hits the ground, it has kinetic energy of $\frac{1}{2}mv^2 = 0.5 \times 0.1 \times 64 = 3.2J$. Therefore, 0.8J of work has been done against resistive forces.

Question 43: E

Take the derivative using the chain rule to get $\frac{dy}{dx} = 6 \times 3 \times (2 + 3x)^5$. Using the binomial expansion, the x^3 term is $18 \times \frac{5!}{3!2!} \times (3x)^3 \times 2^2 = 18 \times 27 \times 4 \times \frac{5 \times 4}{2} = 19440$

Question 44: A

When the stone reaches the top of the cliff on the way down it will have the same kinetic energy as when it is initially released. Therefore, its velocity is $13ms^{-1}$ downwards. Use $s = ut + \frac{1}{2}at^2$ with $a = 10ms^{-2}$ from gravity and $u = 13ms^{-1}$ and choose downwards to be positive. This gives $5t^2 + 13t - 6 = 0$. Factorise to find $(5t - 2)(t + 3) = 0$. Can neglect the negative time so the solution is $t = 2/5 = 0.4s$

Question 45: A

The sum to infinity of a geometric series is $\frac{a}{1-r} = \frac{4}{3}$. As $a = 1$, rearrangement gives $r = \frac{1}{4}$. Therefore, we require solutions to $\sin 2x = \frac{1}{2}$. The solutions to this are

$2x = \frac{\pi}{6} + 2n\pi, \frac{5\pi}{6} + 2n\pi$ for integers n. Divide by 2 for $x = \frac{\pi}{12} + n\pi$ and $x = \frac{5\pi}{12} + n\pi$. Choose $n = 1$ for the desired range to get the answers $x = \frac{13}{12}\pi$ and $x = \frac{17\pi}{12}$.

Question 46: B

The area under the graph is $\frac{1}{2}bh = 0.2 \times 192 = 38.4J$. This is all converted into kinetic energy at the point where the arrow leaves the bow. At the maximum height the arrow has converted all its kinetic energy into gravitational potential energy so $38.4 = mgh = 0.024 \times 10 \times h$. This gives $h = 38.4/0.24 = 160m$. The answer is B.

Question 47: C

The second term is $u_2 = 2p + 3$, the third term is $u_3 = 2p^2 + 3p + 3$ and $u_4 = 2p^3 + 3p^2 + 3p + 3 = -7$. This gives a cubic $2p^3 + 3p^2 + 3p + 10 = 0$. A solution of this equation is $p = -2$. Factorising the cubic gives $(p + 2)(p^2 - p + 5) = 0$. The quadratic factor does not have any real roots, so the only solution is $p = -2$. This gives $u_2 = -1$ and $u_3 = 5$. Summing the four terms gives $2 - 1 + 5 - 7 = -1$

Question 48: H

Resolve the gravitational force parallel and perpendicular to the plane at 20 degrees. The gravitational force down the plane is $mg \sin 20$ and the force into the plane is $mg \cos 20$. The gravitational force is balanced by the resistive force $F = \mu N$ with μ a constant coefficient of friction. As the forces are balanced in the first scenario $\mu mg \cos 20 = mg \sin 20$. This gives $\mu = \tan 20$. Now consider the forces when the surface is at 25 degrees. The gravitational force down the plane is now $mg \sin 25$ and the force into the plane is $mg \cos 25$. The net force down the plane is $mg \sin 25 - \mu mg \cos 25$. Using $F = ma$, the acceleration down the plane is $g(\sin 25 - \cos 25 \tan 20)$.

Question 49: A

The numerator is zero when $x = 1$ so it can be factorised to give $(x - 1)(x^2 - 5x + 4x) = (x - 1)(x - 4)(x - 1)$. The inequality can now be written $\frac{(x-1)^2(x-4)}{x} > 0$. The graph $\frac{(x-1)^2(x-4)}{x} = 0$ crosses the x axis at $x = 4$ but does not at $x = 1$. This is because it is a repeated root. When x is a large positive or negative number the numerator is dominated by the x^3 term so the inequality become $\frac{x^3}{x} = x^2 > 0$ which is true. There is also an asymptote at $x = 0$. Near this point the numerator is dominated by the constant term, so it reduces to $\frac{-4}{x} > 0$. This is true when x is negative and false for positive x. Summarising this the inequality is satisfied when x is negative; between 0 and 1 the function is negative and at $x = 1$ the function is zero. Between 1 and 4 the function is negative and beyond 4 it becomes positive again. Therefore, A is the correct answer.

Question 50: B

As the suitcase is not on the point of slipping we can only use an inequality to describe the resistive force in terms of the coefficient of friction $F > \mu R$. But as the suitcase is travelling at a constant speed the net force on it must be zero. The frictional force must balance the component of the gravitational force that acts down the plane. This is $mg \sin \theta$

Question 51: H

To stretch a graph by scale factor $\frac{1}{2}$ replace all occurrences of x with 2x.

To then shift by $-\pi/4$ in the x-direction replace x with $x + \frac{\pi}{4}$. This gives

$$y = \sin 2\left(x + \frac{\pi}{4}\right) = y = \sin\left(2x + \frac{\pi}{2}\right)$$

Question 52: B

The initial momentum of the ball is $-mu$. The change in momentum is given by the area under the Force-Time graph which is $F(t_2 - t_1)$. The final momentum is $F(t_2 - t_1) - mu$.

Question 53: B

For a straight line, $y = mx + c$, another line described by $y = gx + d$ is perpendicular if $m = \frac{-1}{g}$. Using the gradients from the two lines gives $-1 = (2p^2 - p)(p - 2)$. Expand this to get $2p^3 - 4p^2 - p^2 + 2p = -1$. Collecting the terms gives $2p^3 - 5p^2 + 2p + 1 = 0$. This has a root at **p = 1.** It can be factorised to $(p - 1)(2p^2 - 3p - 1) = 0$. The quadratic factor has solutions $p = \frac{3 \pm \sqrt{17}}{4}$. Take the positive root for the larger solution and take $\sqrt{17} \approx 4$. This gives $p \approx 7/4 = 1.75$

Question 54: A

After hitting the ground for the first time the ball has momentum $P = mv$. It takes 0.8s for the ball to reach the ground again at which point it has an equal velocity but in the opposite direction and momentum $P = -mv$. Use $\Delta P = F\Delta t$ with $F = mg$. This gives $v = \frac{mg \times 0.8}{2m} = 4\text{ms}^{-1}$. The kinetic energy of the ball after the first bounce is $\frac{1}{2}mv^2 = 8m = mgh$ as it is all converted into gravitational energy. This gives $h = \frac{8m}{10m} = 0.8$.

Section 2

Question 1a: B

The acceleration of the ball is a = 0.4t. The velocity is the integral of this v = $0.2t^2$. At t = 0.5s the velocity should be v = $0.05ms^{-1}$ so the correct answer is B.

Question 1b: A

The ball is initially at a distance d. Take downwards as positive and the initial displacement of the ball is −d. The displacement is the integral of the velocity with respect to time which is s = $0.2t^3/3 - d$ using the condition at t = 0. The ball hits the floor when s = 0 giving $t^3 = 15d$. Therefore, t = $(15d)^{1/3}$

Question 1c: A

As the gravitational force continues to increase the ball will require more energy to reach the same height. However, this does not happen, and the ball will bounce less high each time meaning only P is correct.

Question 1d: C

The unit of mass is kg and the unit of acceleration is ms^{-2}. Therefore, multiplying these gives a unit of force: $kgms^{-2}$

Question 1e: E

D is a force so must have units $\mathbf{kgms^{-2}}$. The units of $\mathbf{v^2}$ are m^2s^{-2}. The units of area and density are m^2 and kgm^{-3} respectively. Multiplying these together gives $kgm^2m^2m^{-3}s^{-2} = kgms^{-2}$. This has the same units as a force so X does not have units.

Question 2a: C

A resistor obeys Ohm's law: $V = IR$, so current and voltage are directly proportional to each other. This means the resistor must correspond to the straight line through the origin which is device Y.

Question 2b: B

The filament lamp is 9W at 6V. Use the equation $P = IV$ to find which device has the correct current at this voltage. The current at 6V should be $I = 9/6 = 1.5A$. Therefore, the filament lamp is device X.

Question 2c: D

The resistance of the filament lamp at 6V is $R = V/I = 4\Omega$. The resistor has a resistance of 8Ω. As the resistances are in parallel they add in reciprocal: $\frac{1}{R_T} = \frac{1}{R_1} + \frac{1}{R_2}$. This gives $R_T = \frac{R_1 R_2}{R_1 + R_2} = 32/12 = 8/3\ \Omega$. The current drawn from the supply is then $I = V/R = 18/8 = 2.25A$

Question 2d: C

The same current flows through two components that are in series, so C must be correct.

Question 2e: B

The currents through W and Y are the same and the voltages across the components sum to give 6. Choose a current at which $V_W + V_Y = 6$. This is when $I = 0.5A$, $V_W = 2V$ and $V_Y = 4V$. The power dissipated by W is $P = IV = 1W$

Question 3a: D

Hooke's law behaviour gives a straight line through the origin which is seen up to point P. The strain is given by $\frac{x}{L}$ where L is the natural length of the cord. At the point Q where the cord fractures this is $\frac{0.05}{0.5} = 0.1$

Question 3b: C

The work done is equal to the area under the graph. Each box of 10N by 0.01m is 0.1J of energy. There are approximately six of these under the graph, so the work done is 0.6J

Question 3c: B

As the cord is half the length it can only be stretch half as much before it fractures. This means the maximum work that can be done on the cord is $\frac{U}{2}$. When the catapult is fired this energy is converted into kinetic energy giving $\frac{1}{2}mv^2 = U/2$. Therefore, $V_{max} = \sqrt{U/m}$

Question 3d: D

Now there are two cords twice as much work can be done as in the previous part so $\frac{1}{2}mv^2 = U$. This gives $v = \sqrt{2U/m} = \sqrt{2}V_{max}$

Question 4a: B

The rays from the two slits will positively interfere when the distances travel differ by a whole number of wavelengths. The rays will destructively interfere when the path lengths differ by a half-integer number of wavelengths. Positive interference corresponds to the brightest parts whereas destructive interference gives a minimum. For a diffraction pattern the light from the two slits must have the same frequency i.e. they must be coherent.

Question 4b: D

The introduction of the material will cause the phase of the light from one slit to be shifted. This means the points where the two rays are in phase will move causing the diffraction pattern to shift in the y direction.

Question 4c: E

Radio waves are an EM wave, so they travel at the speed of light. Use $c = f\lambda$ to find the wavelength of the signal: $\lambda = c/f = \frac{3 \times 10^8}{6 \times 10^8} = 0.5$m. This means the separation of the two transmitters is of comparable size to the wavelength, so diffraction will be significant. Therefore, E is the correct answer.

END OF PAPER

2018

Section 1

Question 1: E

It is easiest to present the information from the question in a table:

	Men	Women	**Total**
Passed 1st attempt			*167*
Failed 1st attempt		*143*	
Total	*300*	*200*	

The number of women who passed their first attempt can be calculated as total women – women who failed $= 200 - 143 = 57$

The number of men who passed their first attempt is total people who passed 1st attempt – women who passed 1st attempt $= 167 - 57 = \textbf{110}$

Question 2: B

Alpha particles: atomic number (Z) = 2, mass number (A) = 4

Beta particles: atomic number (Z) = -1, mass number (A) = 0

Particles released: $5\alpha + 2\beta$: $\Delta Z = 5(2) + 2(-1) = \textbf{8}$

Question 3: B

It may help to draw a diagram:

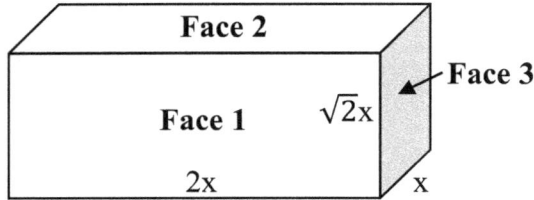

Volume $= (2x)(\sqrt{2}x)(x) = 2\sqrt{2}x^3$

There are 6 faces in total, but 3 different proportions, so surface area is:

$$2(2x)(\sqrt{2}x) + 2(2x)(x) + 2(\sqrt{2}x)(x) = 2x^2(2\sqrt{2} + 2 + \sqrt{2})$$

Volume $= 2 \times$ S.A

$2\sqrt{2}x^3 = 4x^2(3\sqrt{2} + 2)$

$2\sqrt{2}x^3 - 12\sqrt{2}x^2 - 8x^2 = 0$ (factorise out x^2)

It is possible to disregard the $x^2 = 0$ soln. *in this case* because this is the solution for the cuboid having no dimensions, no volume and no surface area. It is not always possible to disregard this solution – take care not to 'divide' by x without good reason, as it may cause solutions to be 'lost'.

$2\sqrt{2}x - 12\sqrt{2} - 8 = 0$

Rearranging for x gives: $x = \frac{12\sqrt{2} + 8}{2\sqrt{2}} = 6 + \frac{8}{2\sqrt{2}} = 6 + 2\sqrt{2}$

Question 4: B

Combining resistors in series using $R_{total} = R_1 + R_2 + R_3 = 30\Omega$

Calculate terminal voltage using $V = IR$. $V = 0.2(30) = 6V$

After removal of resistor R₃: $R_{total} = 27\Omega$

Calculate new current using $I = \frac{V}{R}$. $I = \frac{6}{27} = \frac{2}{9} \approx 0.22$

Question 5: H

Find gradient of line joining the two points in terms of p using $m = \frac{\Delta y}{\Delta x}$:

$$m = \frac{2p - (p-1)}{(1-p) - p} = \frac{p+1}{1-2p}$$

Rearrange the equation of the line to $y = mx + c$ form, to find m:

$$y = -\frac{2}{3}x - \frac{1}{3} \qquad m = -\frac{2}{3}$$

'Parallel' means the two gradients are equal: $\frac{p+1}{1-2p} = -\frac{2}{3}$

Rearranging for p gives: $3(p+1) = -2(1-2p)$

$3p + 3 = -2 + 4p$

$p = 5$

Question 6: G

UV has higher frequency, lower wavelength than visible light.

Convert minimum wavelength of visible light to frequency using $f = \frac{c}{\lambda}$.

$$f = \frac{3 \times 10^8}{400 \times 10^{-9}} = 7.5 \times 10^{14} Hz$$

This is the minimum frequency of UV, as it is the boundary between UV and visible light.

Question 7: D

Let $QR = x$. Draw a labelled diagram:

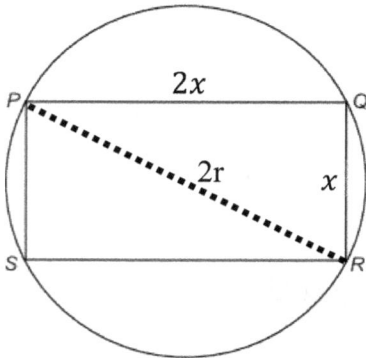

Use area to find 'x': $96 = (2x)(x)$

therefore $x = \sqrt{\frac{96}{2}} = \sqrt{48} = 4\sqrt{3}$

The Pythagorean theorem gives

$$(2r)^2 = (2x)^2 + (x)^2$$

$$4r^2 = \left(8\sqrt{3}\right)^2 + \left(4\sqrt{3}\right)^2 = 240$$

Rearranging for 'r' gives: $r = \sqrt{60} = 2\sqrt{15}$

Question 8: E

Energy transferred (J) = Power (W) x time (s):

$100 \times (10 \times 60) = 60,000$J total energy transferred.

Efficiency of 5% means 95% of the energy is wasted.

Energy wasted $= 0.95 \times 60,000 = 57,000$J.

Question 9: E

If the ratio of the heights is $4:5$, the ratio of volumes is $4^3:5^3$.

Multiplying 320 by a ratio of $64:125$ gives $\frac{320}{64}(125) = 625$cm^3.

Question 10: C

Electrical energy transferred (J) = current (A) x voltage (V) x time (s)

$= 1250 \times 400 \times 4 = 2,000,000$J

Efficiency of 45% means 45% of the energy input is converted to useful energy (kinetic) output.

Kinetic energy $= 0.45 \times 2,000,000 = 900,000$J

Question 11: C

Draw a diagram, labelling the y-intercepts.

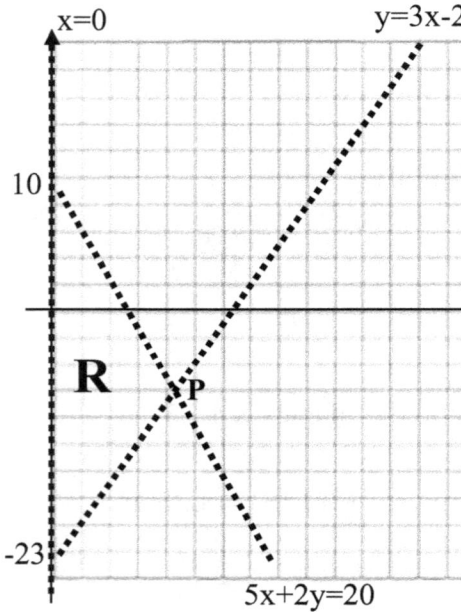

Find P by substituting y=3x-23 into 5x+2y=20:

$$5x + 2(3x - 23) = 20$$

Rearranging give x=6 at P.

Use $area = \frac{1}{2} base \times height$ with base equal to the distance along the y-axis and height equal to the x-coordinate of P.

$$Area = \frac{1}{2}(33)(6) = 99$$

Question 12: A

$$momentum = mass \times velocity \text{ and } K.E. = \frac{1}{2} \times mass \times velocity^2$$

Divide KE by momentum to find velocity: $\frac{K.E.}{momentum} = \frac{\frac{1}{2}mv^2}{mv} = \frac{1}{2}v$

$\frac{1}{2}v = \frac{96}{24}$ therefore $v = 8$m/s

Sub back into momentum or KE equations to find mass: $m = \frac{24}{8} = 3$kg

Question 13: E

Length scale of $1:40$ means volume scale of $1^3:40^3$.

Convert mass of full-sized pillar into cm^3: $12\pi \times 10^6$

Volume of model is therefore $\frac{full-sized\ pillar}{40^3} = \frac{12\pi \times 10^6}{64 \times 10^3} = \frac{3\pi \times 10^3}{16}$

Mass = density x volume $= \frac{4}{3} \cdot \frac{3\pi \times 10^3}{16} = \frac{1000\pi}{4} = 250\pi$

Question 14: A

The background count-rate is the value the count tends towards once the isotope has decayed – in this case 20cpm.

The half-life is calculated as the time taken for the isotope to fall to half the previous value, but an adjustment must be made to remove the background count rate. Therefore the initial reading of 120 on the graph corresponds to an activity of 100cpm for the sample, and so the half-life should be measured to a sample activity of 50cpm (70 on the graph). This gives 40 seconds.

Question 15: D

The internal angles in a regular pentagon are all 108°. RSU is equilateral so ∠RSU=60°. Therefore ∠UST is 48°. As RSU is equilateral, SU=RS. All sides are the same length in a regular pentagon so SU=ST and therefore STU is isosceles. This means ∠STU can be found by:

$$\angle STU = \frac{1}{2}(180 - 48) = 66°$$

Question 16: C

Resultant force down = weight – air resistance = mass x acceleration

$10m - 12 = 2m$ therefore $m = 1.5$kg

Question 17: C

The original price is p. An increase of 125% gives a price of 2.25p. A decrease of 40% gives a price of (0.6)(2.25p) = 1.35p = q.

Question 18: E

A frequency of 10Hz means a particle in the rope completes 10 full cycles per second, or 200 full cycles in 20 seconds.

An amplitude of 4 means the particle travels a distance of 16cm for each full cycle (rest-top-rest-bottom-rest is a full cycle).

The total distance travelled is therefore 20 x 16 = 3200cm.

The speed is unnecessary and is included to obscure the solution!

Question 19: E

Draw a diagram. Arrows show North (N). Using 'C-angles' (interior angles)

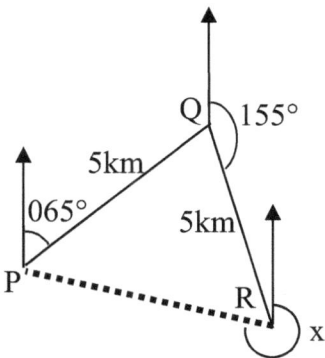

gives $\angle PQN=180-65=115°$

and $\angle QRN=180-155=25°$

At point Q the angle inside the

triangle is 360-155-115=90°

Because PQR is isosceles, once the interior angle at Q is know the angles QPR and QRP can be calculated as $\frac{1}{2}(180-90)=45°$

Finally, the bearing at R can be calculated as 360-25-45=290°.

Question 20: C

Overall equation: $mass\ balance\ reading = m_{cyl} + \rho_{fluid}V$

Forming equation for addition of water: $290 = m_{cyl} + V$

Form equation for addition of oil: $270 = m_{cyl} + 0.9V$

Solving as simultaneous equations gives: $20 = 0.1V$ so $V = 200\text{cm}^3$

Substituting this back into either equation gives $m_{cyl} = 90\text{g}$

Question 21: C

The alternate segment theorem means $\angle PST = \angle PQS = 75°$.

Because the angle between a tangent and a radius is a right angle, $\angle PSO = 90\text{-}75=15°$.

As PQS is isosceles $\angle QSP = \frac{1}{2}(180 - 75) = 52.5°$

$\angle QSO$ is therefore 52.5-15=37.5°.

Question 22: B

At terminal velocity (the straight-line/constant-gradient section of the graph) weight downwards = air restance upwards, giving:

$1000 = kv^2$ so $k = \frac{1000}{v^2}$

The velocity is the gradient of the distance-time graph, and is found to be 50m/s. Therefore $k = \frac{1000}{50^2} = 0.4$.

Question 23: C

From the question: $h = b + 3$ and $area = \frac{hb}{2} = 14$

Combine as simultaneous equations to get $\frac{b(b+3)}{2} = 14$

This gives the quadratic $b^2 + 3b - 28 = 0$, the only appropriate solution of which is $b = 4$cm. Therefore $h = 7$cm.

Use Pythagoras on half the triangle: $s^2 = h^2 + \left(\frac{b}{2}\right)^2 = (b+3)^2 + \left(\frac{b}{2}\right)^2$

Combining terms gives: $s^2 = b^2 + 6b + 9 + \frac{1}{4}b^2 = \frac{5}{4}(4^2) + 6(4) + 9$

$s^2 = 53$ and so s must lie between 7 and 8.

Question 24: H

Nucleons in = 235 (Uranium) + 1 (thermal neutron) = 236

Nucleons out = 236 = 88 (Bromine) + 145 (Lanthanum) + ? (neutrons)

Rearrange to find the number of neutrons out as 236-88-145=3.

Protons in = 92 (Uranium)

Protons out = 92 = 35 (Bromine) + ? (Lanthanum)

Rearrange to find atomic number of Lanthanum is 57.

Beta decay involves a neutron turning into a proton and an electron, therefore when when Lanthanum decays the proton number goes from 57 to 58.

Question 25: G

For the 1st term (n=1): $2 = p(1^2) + q = p + q$

For the 2nd term (n=2): $17 = p(2^2) + q = 4p + q$

Solve as simultaneous equations (subtract) to give: $3p = 15$ so $p = 5$ and $q = -3$.

Therefore $\frac{p-q}{p+q} = \frac{5-(-3)}{5+(-3)} = \frac{8}{2} = 4$

Question 26: C

Let the resistance of X, Y and Z = R.

Create an expression for the total resistance of X and Y: $\frac{1}{R_{total}} = \frac{1}{R} + \frac{1}{R}$

$R_{total} = \frac{R}{2}$

The circuit can therefore be considered as a resistor, Z, of resistance R, in series with a resistor, 'XY', of resistance R/2.

The power dissipated is proportional to the resistance, and therefore two-thirds of the power (12W) will be dissipated through Z, with the other third (6W) dissipated in the combination of X and Y.

The power will split between X and Y equally (3W each).

Question 27: B

The number of red sweets taken will always be exactly 1, because at the point that the child has taken one red sweet, they stop taking any more. Therefore, the options which give more green sweets than red are: 2, 3, 4, 5, or 6 green sweets followed by a red.

Considering the opposite scenarios (fewer green than red) reduces the number of calculations required:

The chance of the child picking a red sweet on their first pick is ½.

The chance of the child picking one green and then red is $\left(\frac{1}{2}\right)\left(\frac{6}{11}\right) = \frac{3}{11}$.

All remaining scenarios give more green than red, so the probability can be worked out as $1 - \frac{1}{2} - \frac{3}{11} = \frac{5}{22}$.

Question 28: A

Distance = speed x time so as wave is travelling at the same speed in all directions, and the time taken to detection is the same for X and Y the distance between the epicentre and X must equal the distance between the epicentre and Y. This doesn't mean that '1' must be true, however, as the epicentre could be equidistant but lie to one side of XY, rather than on the line.

'2' isn't necessarily true, as Z needn't be equidistant from X and Y. For example the scenario below would give the detection pattern seen:

X   Epicentre   Y Z

'3' also needn't be true in all cases. The distance travelled between detection by X or Y, and Z is $distance = speed \times time = 4(60) = 240km$

However, as shown in the diagram, while Z must be no more than 240km from one of X or Y – it could be considerably further from the other (if it was on the opposite side of the epicentre).

Question 29: A

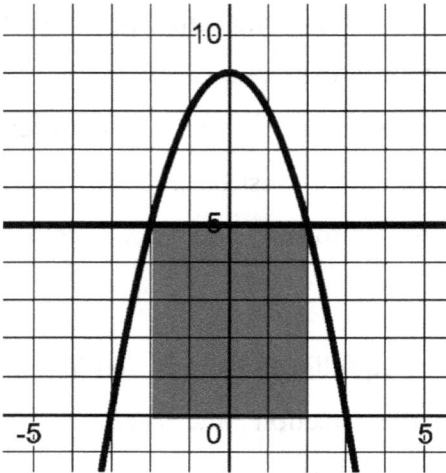

Draw a diagram.

The area between C and L is equal to the area below C (between the two points of intersection) minus the rectangular area below L.

Find the points of intersection:

$9 - x^2 = 5$ gives $x = \pm 2$

Find the area below C by integrating:

$$\int_{-2}^{2} 9 - x^2 \, dx = \left[9x - \frac{x^3}{3} \right]_{-2}^{2} = \left[9(2) - \frac{2^3}{3} \right] - \left[9(-2) - \frac{-2^3}{3} \right] = \frac{92}{3}$$

The area below L is base x height = 4 x 5 = 20

The area between C and L is therefore $\frac{92}{3} - 20 = \frac{32}{3}$.

Question 30: E

SUVAT: $s = 1{,}600$m, $u = 0$m/s, $v = 80$m/s, $a = ?$

Use $v^2 = u^2 + 2as$: $80^2 = 2(1600)a$. Therefore $a = \frac{6400}{3200} = 2$m/s^2.

Question 31: D

Rearranging gives: $2 \sin^3 \theta - \sin \theta = 0$

Factorise (dividing would lose solutions): $sin\theta (2 \sin^2 \theta - 1) = 0$

Solutions are $sin\theta = 0$ or $sin\theta = \pm \frac{1}{\sqrt{2}}$

Considering the graph of $y = sin\theta$ (overlayed with $y = 0$, $y = \frac{1}{\sqrt{2}}$, and $y = -\frac{1}{\sqrt{2}}$) shows 5 solutions in the interval $-\frac{\pi}{2} \leq \theta \leq \pi$:

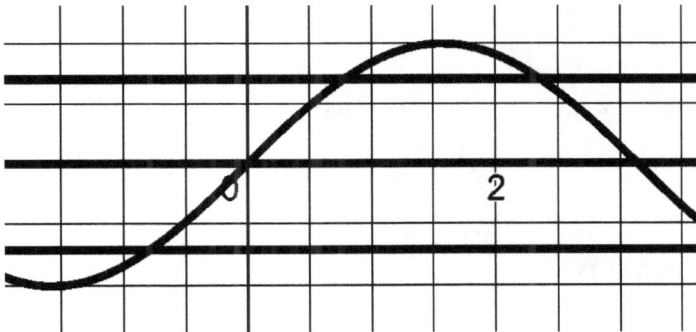

Question 32: F

A: Not necessarily true – balanced overall doesn't mean 'equal'. B: There will likely be more than 3: normal reaction, friction, weight, driving force. C: Not true – the driving force up the slope will be balanced by both friction and the component of the weight parallel to the slope. D: Not true – mass is straight down, contact is perpendicular to plane. E – Not enough information to know. F – This must be correct. Constant speed means no overall resultant force (Newton's First Law).

Question 33: C

The tangent to the curve has gradient found by differentiation:

$$\frac{dy}{dx} = 6x - 2$$

The gradient of the curve is equal to the gradient of the tangent, so:

$6x - 2 = 1$ therefore $x = \frac{1}{2}$

The y-coordinate is found by substituting x back into the equation of the

curve: $y = 3\left(\frac{1}{2}\right)^2 - 2\left(\frac{1}{2}\right) + 1 = \frac{3}{4}$

The value of k can then be found as $\frac{3}{4} = \frac{1}{2} + k$, giving $k = \frac{1}{4}$.

Question 34: D

All blocks have the same acceleration, calculated using Newton's 2nd Law (N2L): $30 = (3 + 4 + 6 + 2)a$ giving $a = 2\text{m/s}^2$.

Considering N2L on 'Z' alone: $30 - T_1 = 2(2)$ gives $T_1 = 26\text{N}$

Then considering N2L on 'Y' alone: $T_1 - T_2 = 6(2)$

Therefore $26 - T_2 = 12$, so $T_2 = 14\text{N}$.

Question 35: D

Area of a sector $= \frac{1}{2}r^2\theta$ and Arc length $= r\theta$

$Area\ (S) = \frac{1}{2}r^2\theta = 10\pi$ and $Area\ (T) = \frac{1}{2}r^2\left(\theta + \frac{\pi}{20}\right) = \frac{25}{2}\pi$

Solving simultaneously (subtracting Area (S) from Area (T)) gives

$\frac{1}{2}r^2\left(\frac{\pi}{20}\right) = \frac{5}{2}\pi$ therefore r can be calculated as 10cm.

Substituting back into the expression for Area (S) gives $\theta = \frac{\pi}{5}$.

Arc length (T) $= 10\left(\frac{\pi}{5} + \frac{\pi}{20}\right) = \frac{10\pi}{4} = \frac{5}{2}\pi$

Question 36: F

Let contact force at X $= F_x$.

Take moments around Y: $10g(2) + 40g(3) = F_x(4)$

Solve for $F_x = 35g = 350$N.

Question 37: E

n^{th} term of A.P $= a + (n - 1)d$

13th term $= 6$ x 1st term: $a + 12d = 6[a]$

11th term $= 2$ x 5th term $- 1$: $a + 10d = 2[a + 4d] - 1$

Combine terms: $5a = 12d$ and $a - 1 = 2d$

Solve simultaneously: $5 = 2d$ so $d = \frac{5}{2}$ and $a = 6$

3rd term $= a + 2d = 6 + 2\left(\frac{5}{2}\right) = 11$

Question 38: B

$Initial\ GPE = Final\ KE + Work\ done\ against\ friction$

$mglsin\theta = KE + kmglsin\theta$ therefore $KE = mglsin\theta(l - k)$

As a proportion of the initial GPE, $mglsin\theta$:

$\frac{KE}{GPE} = \frac{mglsin\theta(l-k)}{mglsin\theta} = 1 - k$

Question 39: B

n^{th} term of G.P $= ar^{n-1}$ where r is a constant ratio, therefore:

1st term: $a = p - 2$

2nd term: $ar = 2p + 2$ so $r = \frac{2p+2}{p-2}$

3rd term: $ar^2 = 5p + 4$ so $r = \frac{5p+14}{2p+2}$

Solving simultaneously by setting the two expressions for r equal to each other gives p = 8 or p = -4.

'p' must equal 8 as the question states the terms are all greater than 0. 'r' can be calculated as $\frac{2(8)+2}{8-2} = 3$ and 'a' as $8 - 2 = 6$.

The 5th term is therefore $ar^4 = 6(3^4) = 486$.

Question 40: D

Change in momentum = Impulse = Force x time

Change in momentum of X when force is applied = 5 x 3 = 15 kgm/s

$\Delta mv = m(v_f - v_1)$ gives $15 = 2(v_f - 4.5)$ and therefore $v_f = 12$m/s

Conservation of momentum:

Momentum of X+Y before = Momentum of XY after

$2(12) + 3(0) = (3 + 2)v_{combined}$ which can be rearranged to give the speed of the combined 'XY' as $\frac{24}{5}$m/s.

Question 41: C

In general $\log(a) + \log(b) = \log(ab)$

$$\log_2\left(\frac{5}{4}\right) + \log_2\left(\frac{6}{7}\right) + \cdots + \log_2\left(\frac{64}{63}\right) = \log_2\left(\frac{(5)(6)\ldots(64)}{(4)(5)\ldots(63)}\right)$$

As an aside: $n! = n \times (n-1) \times (n-2) \ldots \times 3 \times 2 \times 1$.

In this case, not all terms are included, so instead of 64! the expression for the numerator is $\frac{64!}{4!}$. Similarly, instead of 63! the expression for the denominator is $\frac{63!}{3!}$.

The expression is therefore equal to $\log_2\left(\frac{64!}{4!} \cdot \frac{3!}{63!}\right)$ which simplifies to

$$\log_2\left(\frac{64}{4}\right) = \log_2(16) = 4$$

Question 42: G

Initial energy $= KE + GPE = \frac{1}{2}(0.2)(4^2) + 0.2(10)(0.45) = 2.5J$.

Energy after bounce $= KE = \frac{1}{2}(0.2)(2^2) = 0.4J$

Energy lost in bounce $= 2.5 - 0.4 = 2.1J$.

Question 43: C

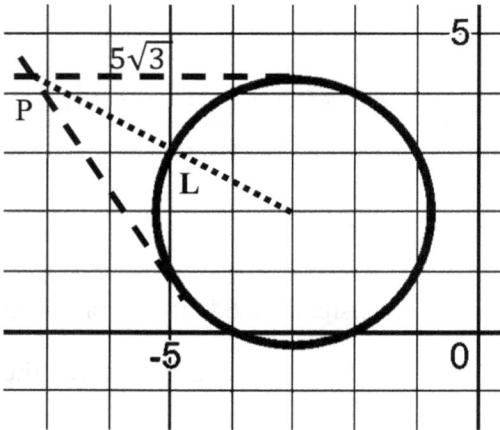

Draw a diagram.

Consider the simplest case, when the tangent is horizontal (or vertical).

The co-ordinates of P would therefore be:

$x = -3 - 5\sqrt{3}$ and $y = 2 + \sqrt{5}$.

The shortest distance between the point P and the circle can then be generalised as the length of the line marked 'L' minus the radius.

$$Shortest\ distance = \sqrt{(x_p - x_c)^2 + (y_p - y_c)^2} - r$$ where the point P

has coordinates (x_p, y_p) and the circle has centre (x_c, y_c).

Substituting in the values above gives *shortest distance* $= 3\sqrt{5}$.

Question 44: F

(Momentum of X+ Momentum of Y) before = Momentum of 'XY' after

After: XY both have velocity '-v', and therefore momentum = -3mv

Before: X has momentum 'mv'

Rearrange to find the momentum of Y before = -3mv – mv = -4mv

Speed of Y before is therefore -2v.

Question 45: D

Distance AC = $\sqrt{(u-2)^2 + (-1-(-3))^2} = \sqrt{u^2 - 4u + 8}$

Distance AB = $\sqrt{(3-2)^2 + (1-(-3))^2} = \sqrt{1+16} = \sqrt{17}$

AC=2AB: $\sqrt{u^2 - 4u + 8} = 2\sqrt{17}$

Square both sides and rearrange: $u^2 - 4u - 60 = 0$

Factorising gives: $(u-10)(u+6) = 0$ so $u = 10$ or $u = -6$.

Question 46: A

Resolve horizontally: $P = T sin30$ $\qquad\qquad$ $(sin(30) = \frac{1}{2})$

Question 47: D

'Brute force' method:

Binomial expansion: $(a + x)^n = a^n + na^{n-1}x + \frac{n(n-1)}{2!}a^{n-2}x^2 + \cdots$

Expansion of $(1 - 2x)^5 = 1 + 5(-2x) + \frac{5(4)}{2}(-2x)^2 + \frac{5(4)(3)}{(3)(2)}(-2x)^3$

$= 1 - 10x + 40x^2 - 80x^3 + \cdots$

Expansion of $(1 + 2x)^5 = 1 + 5(2x) + \frac{5(4)}{2}(2x)^2 + \frac{5(4)(3)}{(3)(2)}(2x)^3 + \cdots$

$= 1 + 10x + 40x^2 + 80x^3 + \cdots$

When multiplied together the x^3 terms would be:

$1(80x^3) - 10x(40x^2) + 40x^2(10x) - 80x^3(1) = 0$

Alternative 'clever' method:

$(1 - 2x)^5(1 + 2x)^5 = [(1 - 2x)(1 + 2x)]^5 = [1 - 4x^2]^5$

Because the bracket to be expanded contains only an x^2 term, there will be no terms in the expansion with x raised to an odd power.

Question 48: A

Although this looks like SHM, the angle is not small enough for the equations to be appropriate.

Find the vertical distance between 'pivot' and pendulum at the 'amplitude position' using Pythagoras: $\sqrt{50^2 - 30^2} = 40$cm.

This means between 'amplitude' and 'equilibrium' the bob loses '10cm' of GPE: $0.01(10)(0.1) = 0.01$

It will gain KE = GPE lost, and therefore $0.01 = \frac{1}{2}mv^2$ can be rearranged to give $v = \sqrt{\frac{0.02}{0.01}} = \sqrt{2}$m/s.

Question 49: E

$$\int_0^2 x^m dx = \left[\frac{x^{m+1}}{m+1}\right]_0^2 = \left[\frac{2^{m+1}}{m+1}\right] = \frac{16\sqrt{2}}{7}$$

$$\int_0^2 x^{m+1} dx = \left[\frac{x^{m+2}}{m+2}\right]_0^2 = \left[\frac{2^{m+2}}{m+2}\right] = \frac{32\sqrt{2}}{9}$$

$\left[\frac{2^{m+2}}{m+2}\right] = \left[\frac{2^{m+1}}{m+1}\right]\left[\frac{2(m+1)}{m+2}\right]$ giving $\frac{32\sqrt{2}}{9} = \frac{16\sqrt{2}}{7}\left[\frac{2(m+1)}{m+2}\right]$

Which can be rearranged for $\frac{1}{9} = \frac{m+1}{7(m+2)}$ giving $m = \frac{5}{2}$.

Question 50: E

Total momentum (to the right) before $= mu$

Total momentum (to the right) after $= MU_M - mv$

Setting before and after equal gives $MU_M - mv = mu$ which can be rearranged for $U_M = \frac{mu+mv}{M}$.

Question 51: A

$f'(x) = ax + g(x)$ can be integrated between 2 and 4 to give:

$$\int_2^4 f'(x)dx = \int_2^4 ax + g(x)dx$$

$$f(4) - f(2) = \left[\frac{ax^2}{2}\right]_2^4 + \int_2^4 g(x)dx$$

$$18 = \left(\frac{4^2}{2} - \frac{2^2}{2}\right)a + 12 \text{ giving } a = \frac{6}{6} = 1.$$

Question 52: F

Initial GPE = Final KE + Work done against friction where the work done against friction is half the change in GPE. $0.5mgh = \frac{1}{2}mv^2$ gives $0.5(10)h = \frac{1}{2}(10^2)$ which can be solved for $h = 10$m.

Question 53: C

It may help to draw a diagram:

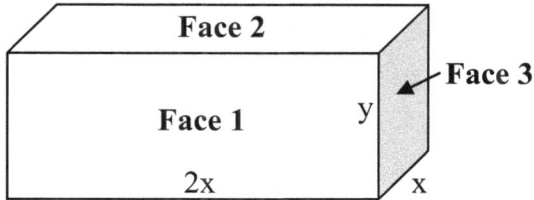

Volume of the cuboid $= 2x^2y = 576$ this can be rearranged to give $y = \frac{288}{x^2}$.

S.A of the cuboid $= 2(2xy + xy + 2x^2) = 4x^2 + 6xy$

Substitute to eliminate y: $S.A = 4x^2 + \frac{1728}{x}$

If this is a maximum value then the differential of the surface area with respect to x (or y) is equal to 0.

$\frac{dSA}{dx} = 8x - \frac{1728}{x^2} = 0$ can be rearranged to find $x = 6$.

When $x = 6$, $y = 8$ and the largest face therefore has surface area $= 12\text{x}8 = 96 \text{ cm}^2$.

Question 54: B

SUVAT (1st object): $s = 0$m, $a = -10$m/s^2, $u = 40$m/s, $t = T$

Use $s = ut + \frac{1}{2}at^2$ for both objects.

For object 1 this gives: $0 = 40T - 5T^2$ which can be solved to give $T = 0$ which isn't a relevant solution, or $T = 8s$.

SUVAT (2nd object): $s = -h$m. $a = -10$m/s^2, $u = 0$m/s, $t = 8 - 2$s

This gives $-h = -5(6^2)$ therefore $h = 180$m.

END OF SECTION

Section 2

Question 1: B

Man: distance = speed x time = 9t

Boy – use SUVAT: $a = 0.8\text{m/s}^2$, $u = 5\text{m}$, $t =$?

$$s = ut + \frac{1}{2}at^2 = 5t + 0.4t^2$$

Set the distance travelled by each equal to each other:

$9t = 5t + 0.4t^2$ can be solved to give $t(0.4t - 4) = 0$

This means the boy and man have the same displacement at t=0 (given in the question) and at t=10s.

Question 2: B

$$\rho_{mixture} = \frac{mass_{mixture}}{volume_{mixture}} = \frac{\rho_p V_p + \rho_Q V_Q}{V_p + V_Q}$$

Question 3: C

The resistance of the combination (WY) would increase if W increased. Considering these as a potential divider would mean that the voltage dropped across WY would increase, and the voltage dropped across X would decrease. As a result, the power dissipated through WY would increase, and so the power dissipated in X would decrease.

The current flowing through Y would increase and so would the power dissipated through Y.

Question 4: A

As there are no external forces acting, the total momentum of the system is 0 at all points. Therefore following the collision, the magnets must have 0 velocity.

Question 5: F

In diagram 2: The buoyancy force of the water on the stone = 1N. Therefore the force of the stone on the water (and by extension the balance) = 1N. The reading will have increased by 100g.

In diagram 3: The stone weighs 300g (as shown in diagram 1), therefore the mass balance reading will have increased by 300g.

Question 6: F

From diagram 1: The wavelength is twice the length of the tube = 1m.

Use $f = \frac{c}{\lambda}$ to find frequency is 1000Hz and $T = \frac{1}{f} = \frac{1}{1000} s$

From diagram 2: X is 1.5 cycles from the origin, therefore it occurs at 0.0015s.

Question 7: A

The collision is elastic so there is no loss in kinetic energy due to the collision.

The mass of the tennis ball can be worked out from KE as $m = \frac{2KE}{v^2} = 0.06kg$

Impulse applied = Force x time = Area under graph $= \frac{2}{1000} \times 1500 = 3Ns$.

Change in momentum = Impulse = mass (final velocity − initial velocity)

Therefore $3 = 0.06(v − (−30))$ so $v = \frac{3}{0.06} − 30 = 20$m/s.

Question 8: G

Let the reaction force at P be F_p, and the reaction force at Q be F_Q.

Take moments around P: $5g(0.2) + 74g(1) + 24g(1) = F_Q(1.5)$

Rearranging gives $F_Q = \frac{1+74+24}{1.5}g = 660\text{N}$.

Question 9: E

Consider the top 'branch' as a potential divider with total voltage 12V. The voltage dropped across the 2 and 3Ω resistors is $12\left(\frac{2+3}{1+2+3}\right) = 10\text{V}$. Therefore the 'top' of the voltmeter is at 2V. Considering the bottom 'branch' similarly gives a voltage drop of 2V meaning the 'bottom' of the voltmeter is at 10V. The voltmeter reading will be the difference.

Question 10: D

Spring constants can be combined (with the opposite rules to resistors).

The springs in parallel therefore have a combined constant of 3k.

This can be combined with the spring in series as $\frac{1}{k_{total}} = \frac{1}{3k} + \frac{1}{k}$ gives $k_{total} = \frac{3k}{4}$. Extension, $x = \frac{F}{k} = \frac{4mg}{3k}$

$$EPE = \frac{1}{2}Fx = \frac{1}{2}mg\left(\frac{4mg}{3k}\right) = \frac{2(mg)^2}{3k}$$

Question 11: B

GPE lost = KE gained + work done against friction

$3.6(10)(1.5sin30) = \frac{1}{2}(3.6)(2^2) + W_f$ can be rearranged to find $W_f = 19.8J$

Rate at which work is done $= \frac{W_f}{time} = W_f.\frac{av.\ speed}{distance} = 19.8\left(\frac{1}{1.5}\right) = 13.2J/s$.

Question 12: D

The total current in the circuit = current through R=20 + current through R=30

$$I_{total} = \frac{4.8}{20} + \frac{4.8}{30} = 0.24 + 0.16 = 0.4A$$

If 6V are produced by the battery, but there are only 4.8 over the resistors, the 'lost volts' dropped due to the internal resistance must be 1.2V.

The internal resistance can be calculated as lost volts divided by circuit current: $r = \frac{1.2}{0.4} = 3\Omega$

Question 13: F

The stick dips into the water every 0.8s, therefore T=0.8s. In the 1second which has passed, the wave has progressed by 1.25 wavelengths. Therefore the difference between wavecrest Q at time t and the wavecrest at time t+1 corresponds to 1.25 wavelengths.

0.25 wavelengths = 1.5cm, so the wavelength is 6cm

Question 14: B

Though counterintuitive, because the surfaces of P and Q are smooth Q is not being held in place by friction. This means that the acceleration of Q due to the tension in the string is equal to the acceleration of P.

Resolve vertically on R: $m_R g = T$

Resolve horizontally on Q: $T = m_Q a$

This gives: $\frac{m_r g}{m_q} = a$.

Question 15: G

Surface area of a cube $= 6l^2 = 96$ therefore $l = 4$cm.

$F = kx$ finds the weight of the cube as $2 \times 10^4 \times 1.6 \times 10^{-4} = 3.2$N

Pressure $= \frac{Force}{Area} = \frac{3.2}{(4 \times 10^{-2})^2} = 2000$N/m^2

Density $= \frac{mass}{volume} = \frac{\frac{W}{g}}{l^3} = \frac{0.32}{(4 \times 10^{-2})^3} = 5{,}000$kg/m^3.

Question 16: A

$$M = density_{copper}(A)(L) + density_{aluminium}(6A)(L) = 9ALd \qquad \text{or}$$

$$A = \frac{M}{9Ld}$$

Find the resistance of copper cylinder as $\frac{\rho L}{A}$. $R_c = \frac{2\rho L}{A}$

Find the resistance of a single aluminium cylinder as $R_a = \frac{3\rho L}{A}$

Combining as resistors in parallel: $\frac{1}{R_{total}} = \frac{A}{2\rho L} + \frac{6A}{3\rho L}$ gives $R_{total} = \frac{2\rho L}{5A}$

Substituting A gives $R_{total} = \frac{2\rho L}{5}\left(\frac{9Ld}{M}\right) = \frac{18\rho L^2 d}{5M}$

Question 17: D

Mass flow rate $= \rho A v = 2,400$kg/s

Mass flow rate must be conserved. This means the velocity of the fluid in the smaller cross-section is $\frac{2400}{800 \times 0.25} = 12$m/s.

$$Force = \frac{\Delta mv}{t} = \frac{m}{t}.\Delta v = 2400(12 - 5) = 16,800N$$

Question 18: B

Consider between release and Q: GPE lost=Work done against friction

Work done = force x distance (in this case the proportion of the circumference)

$mgrcos45 = 2\pi r \left(\frac{135}{360}\right) F$ where 'F' is the constant magnitude of the friction.

Rearranging gives $F = \frac{2\sqrt{2}mg}{3\pi}$.

Considering between release and P: GPE lost = KE gained + work done

$mgr = KE + 2\pi r \left(\frac{90}{360}\right) F$ gives $KE = mgr - \frac{\pi r F}{2}$

The expression for F can then be substituted in.

END OF PAPER

Afterword

Remember that the route to a high score is your approach and practice. Don't fall into the trap that *"you can't prepare for the ENGAA"*– this couldn't be further from the truth. With knowledge of the test, time-saving techniques and plenty of practice you can dramatically boost your score.

Work hard, never give up and do yourself justice.

Good luck!

About UniAdmissions

UniAdmissions is an educational consultancy that specialises in supporting **applications to Medical School and to Oxbridge**.

Every year, we work with hundreds of applicants and schools across the UK. From free resources to our *Ultimate Guide Books* and from intensive courses to bespoke individual tuition – with a team of **300 Expert Tutors** and a proven track record, it's easy to see why UniAdmissions is the **UK's number one admissions company**.

To find out more about our support like intensive **courses** and **tuition**, check out **www.uniadmissions.co.uk/engaa**

www.ingramcontent.com/pod-product-compliance
Lightning Source LLC
Chambersburg PA
CBHW070735220326
41598CB00024BA/3437